DATE			

Student's Guide to Structural Design

S. A. Lavan
Senior Lecturer in Civil Engineering Studies, Westminster College, London

B. G. Fletcher
Head of the Department of Construction, Waltham Forest College, London

Butterworths
London Boston Singapore Sydney Toronto Wellington

PART OF REED INTERNATIONAL P.L.C.

First published 1989

Butterworth International Edition, 1989
ISBN 0 408 05309 7

British Library Cataloguing in Publication Data
Lavan, S. A. (Sean A.)
Student's guide to structural design.
1. Structural engineering. Design
I. Title II. Fletcher, B. G. (Bernard G.)
624.1'771

ISBN 0-408-02171-3

Library of Congress Cataloging in Publication Data applied for

Typeset by KEYTEC, Bridport, Dorset
Printed and bound by Hartnolls Ltd, Bodmin, Cornwall

Preface

The purpose of this book is to provide a basic introduction to the requirements of structural design in reinforced concrete, prestressed concrete, structural steelwork, structural timber and unreinforced masonry as set out in the relevant British Standards Institution codes of practice. It is hoped that by presenting the design procedures in a logical series of stages the reader will find the apparently daunting task of following the requirements of the codes less so. The authors have, where possible, set out the design procedures in a manner that allows the calculations to be carried out without reference to the actual code in question. However the book is not intended to be a complete substitute for the codes. It is hoped that students of structural engineering, civil engineering and architecture, as well as practising engineers who require a basic introduction to the requirements at the limit state design codes, will find this book both helpful and informative.

Our thanks are due to Trevor Fairman, who so expertly prepared the illustrations.

Extracts from British Standards are reproduced with the permission of BSI. Complete copies of the British Standards can be obtained by post from BSI Sales, Linford Wood, Milton Keynes, Bucks. MK14 6LE.

S. A. Lavan
B. G. Fletcher

Contents

1

General

1.1 SI Units

The Système International d'Unités or, in its abbreviated form, SI is an international system of measurement based upon six fundamental units (Table 1.1). These units can be combined in either product or quotient form to derive SI units (Table 1.2). The unit of force, the newton, is derived from the unit of mass through the relationship that force is equal to mass times the gravitational constant of 9·81 m/s^2: e.g.

1000 kg = 1000 × 9·81 kg m/s^2 = 9810 N

A detailed description of the system of SI units is given in BS 3763: 1964, *The International System (SI) Units*. To express magnitudes of a unit, decimal multiples or submultiples are formed using a prefix with the name of the unit (Table 1.3).

Table 1.1 Basic SI units

Quantity	*Name of unit*	*Unit symbol*
Length	metre	m
Mass	kilogram	kg
Amount of substance	mole	mol
Time	second	s
Electric current	ampere	A
Thermodynamic temperature	degree kelvin	K
Luminous intensity	candela	cd

Table 1.2 Some derived SI units

Quantity	*Name of unit*	*Unit symbol*
Force	newton	N
Area	square metre	m^2
Volume	cubic metre	m^3
Density	kilogram per cubic metre	kg/m^3
Pressure and stress	newton per square metre	N/m^2

Table 1.3 Magnitudes of SI units

Multiplication factor		*Prefix*	*Symbol*	*Example*
1 000 000 000	10^9	giga	G	giganewtons (GN)
1 000 000	10^6	mega	M	megawatt (MW)
1 000	10^3	kilo	k	kilometre (km)
100	10^2	hecto	h	These factors are non-preferred and should be avoided
10	10	deca	da	
0·1	10^{-1}	deci	d	
0·01	10^{-2}	centi	C	
0·001	10^{-3}	milli	m	millimetre (mm)
0·000001	10^{-6}	micro	μ	microsecond (μs)

Indices

It will be noted that in Table 1.3 the multiplication factor has been expressed in two ways: as a whole number and as a power of 10 raised by an appropriate index: e.g.

1 000 000 or 10^6

The method of expressing large or very small numbers in terms of the power 10 raised by an index is of considerable use to the structural engineering student. The laws of indices are as follows:

1. 10 = 10× 10 × 10 . . . to occurrences of 10

   e.g. $10^4 = 10 \times 10 \times 10 \times 10 = 10\,000$

   It follows, for example, that 2 340 000 can be rewritten in the convenient form $2{\cdot}34 \times 10^6$.

2. The product of powers.

   $10^x \times 10^y = 10 \times 10 \times 10 \ldots$ to x occurrences of 10 $\times 10 \times 10 \times 10 \ldots$ to y occurrences of 10 $= 10^{x+y}$

   e.g. $10^6 \times 10^8 = 10^{6+8} = 10^{14}$

3. The quotient of powers

   $$\frac{10^x}{10^y} = \frac{10 \times 10 \times 10 \ldots \text{ to } x \text{ indices}}{10 \times 10 \times 10 \ldots \text{ to } y \text{ indices}} = 10^{x-y}$$

   e.g. $\frac{10^{12}}{10^9} = 10^{12-9} = 10^3$

4. The product of indices.

   $(10^x)^y = 10^{xy}$

   e.g. $(10^3)^4 = 10^{3\times4} = 10^{12}$

   $(10^4)^3 = 10^{4\times3} = 10^{12}$

5. 10 raised to the power of zero equals 1, i.e.

   $10^0 = 1$

   because $\frac{10^1}{10^1} = 10^{1-1} = 10^0$

   but $\frac{10^1}{10^1} = 1$

   therefore $10^0 = 1$

6. Reciprocals.

   $\frac{1}{10} = 10^{-1}$

   or in general terms

   $\frac{1}{10^x} = 10^{(-x)}$

7. Other laws.

   (a) $(10a)^x = 10^x \times a^x$

   e.g. $(10 \times 14)^3 = 10^3 \times 14^3$

   $= 10^3 \times 2744$

   $= 10^3 \times (2{\cdot}744 \times 10^3)$

   $= 2{\cdot}744 \times 10^6$

   (b) $\left(\frac{10^x}{10^y}\right)^z = \frac{10^{xz}}{10^{yz}} = 10^{xz-yz}$

   e.g. $\left(\frac{10^2}{10^5}\right)^4 = \frac{10^8}{10^{20}} = 10^{8-20}$

   $= 10^{-12} = \frac{1}{10^{12}}$

   (c) $10^{1/x} = \sqrt[x]{10}$

   e.g. $10^{1/3} = \sqrt[3]{10}$

   (d) $10^{x/y} = \sqrt[y]{10^x}$

   e.g. $10^{3/4} = \sqrt[4]{10^3}$

Compatibility of units

If calculations are to be successful it is imperative that the quantities involved are expressed in compatible units. If, for example, it is necessary to calculate the area of a rectangle whose sides are 1·5 m and 350 mm it is first necessary to express the 1·5 m in millimetres or 350 mm in terms of metres.

The former action would result in an answer where the derived unit would be mm^2 and in the latter case the derived unit would be m^2:

i.e. $1\text{ m} = 10^3\text{ mm}$

hence area $= 15 \times 10^3 \times 350$

$= 525 \times 10^3\text{ mm}^2$

$= 525\,000\text{ mm}^2$

In the latter case:

from $1\text{ m} = 10^3\text{ mm}$

$1\text{ mm} = \frac{1}{10^3}\text{ m} = 10^{-3}\text{ m}$

hence area $= 1{\cdot}5 \times (350 \times 10^{-3})$

$= 525 \times 10^{-3}\text{ m}^2$

$= 0{\cdot}525\text{ m}^2$

To convert kN m to N mm:

$1\text{ kN} = 10^3\text{ N}$

$1\text{ m} = 10^3\text{ mm}$

hence $1\ \text{kN m} = (10^3\ \text{N}) \times (10^3\ \text{mm}) = 10^6\ \text{Nmm}$

To convert N/mm^2 to kN/m^2:

$$1\ \text{N} = \frac{1}{10^3}\ \text{kN} = 10^{-3}\ \text{kN}$$

$$1\ \text{mm} = \frac{1}{10^3}\ \text{m} = 10^{-3}\ \text{m}$$

hence $1\ \text{mm}^2 = (10^{-3}\ \text{m})^2 = 10^{-6}\text{m}^2$

$$\text{hence}\ \frac{\text{N}}{\text{mm}^2} = \frac{10^{-3}\ \text{kN}}{10^{-6}\ \text{m}^2} = 10^{-3+6}\ \text{kN/m}^2 = 10^3\ \text{kN/m}^2$$

SI Notation

The decimal marker is the conventional decimal point, raised from the line, not the comma used in some early British publications and still in use in some metric countries. The full stop on the line is used in typewriting unless the machine has a special character above the line. Values less than unity should always have a zero before the decimal point, so that we write 0·600, for example, *not* ·600. No full stop is used for abbreviations used as symbols for units: for example, the abbreviation for millmetre is 'mm', *not* 'mm.'. Where there is a group of five or more digits to left or right of the decimal point they are grouped in threes with a space, not a comma dividing them; thus:

0·1041	1041·0
0·10411	11041·0
0·104111	111041·1

On drawings, dimensions are always in metres or millimetres, indicated by figures only, with no symbol. A dimension in metres always has three digits after the decimal point; thus:

1·740 0·320 14·000	all metres
14 112·5 2741	all millimetres

1.2 Definitions

Effective length of columns

See 'Slenderness ratio' (page 6).

Elasticity

A material is said to be elastic if it displays the characteristics of reverting to its original shape after an applied force, which has caused a deformation of its shape, is removed.

End fixity of beams

End fixity refers to the degree of restraint against rotation, in the plane of bending, of a beam at its end supports when a force is applied to that beam. The two extreme conditions of end fixity are:

1. Fully fixed end conditions, in which case the ends of the beam can resist rotation and hence bending forces. Bending moments are induced at the supports (Figure 1.1).

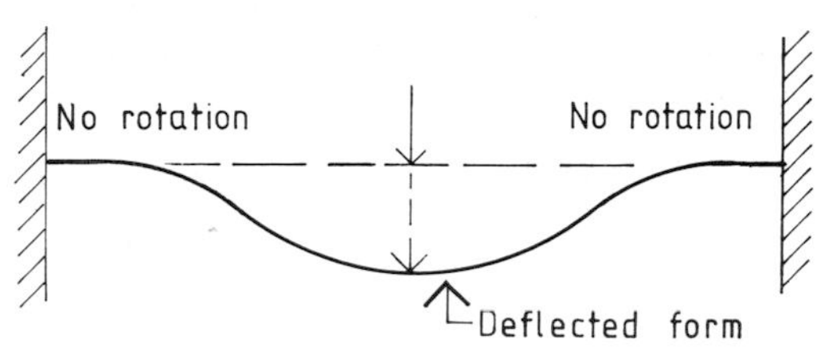

Figure 1.1 Fully fixed end conditions

2. Simply supported end conditions, in which case the ends of the beam are free to rotate and hence do not resist bending forces. A characteristic of a simple support is that it cannot induce a bending moment (Figure 1.2).

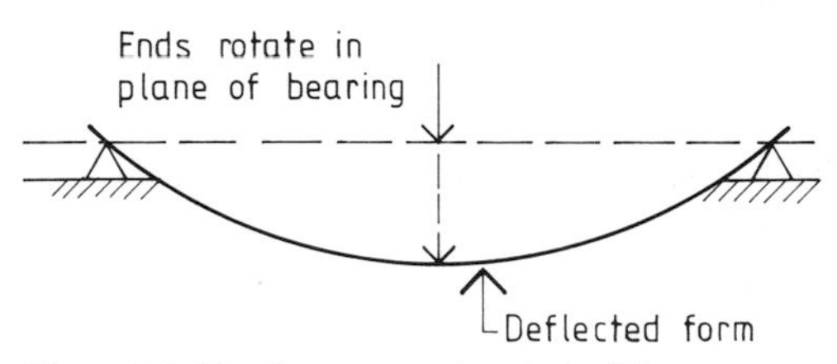

Figure 1.2 Simply supported end conditions

In practice the degree of fixity of the beam supports are somewhere between the two cases described above. It is therefore normal practice when calculating the bending moments to consider a beam as being simply supported unless the end fixity can be determined.

Factor of safety

With regard to the design of structural members, etc., the factor of safety is given by

$$\frac{\text{Ultimate or failing stress}}{\text{Permissible stress}}$$

(see also 'Load factor', below). For materials such

as mild steel which have a definite yield point the factor of safety is sometimes taken as

$$\frac{\text{Yield stress}}{\text{Permissible stress}}$$

In overturning calculations for walls the factor of safety against overturning is:

$$\frac{\text{Stability moment}}{\text{Overturning moment}}$$

For sliding of walls the factor of safety is

$$\frac{\text{Force resisting sliding}}{\text{Force causing sliding}}$$

Force

Force is any cause which produces or tends to produce motion or change of motion in the body on which it acts. It is measured in newtons, and one newton is defined as that force which, when applied to a mass of 1 kg, gives that mass an acceleration of 1 m/s^2.

Hooke's law

This law states that the deformation in an elastic material is proportional to the load on it. As deformation can be described by strain and load can be considered in terms of stress, it can be seen that in an elastic body strain is proportional to stress.

Load factor

The load factor of a member is defined as

$$\frac{\text{Load which would cause failure}}{\text{Design load}}$$

This, of course, gives a margin of safety, but load factor is not the same as factor of safety. Load factor design is based on actual conditions of stress at failure, when stress is not proportional to strain. The factor of safety applied, for example, to beams is usually based on the failing (or yield) stress in tension or compression and the beams are designed assuming elasticity.

Middle-third rule

This rule is usually employed in the design of masonry and mass concrete construction (e.g. foundations, piers and retaining walls) and applies to rectangular cross sections. In order that tensile stresses do not develop in the material the resultant thrust must not be outside the kern shown shaded in Figure 1.3.

Modulus of elasticity or Young's modulus

This refers to the lengthening or shortening of members subjected to a direct force. Since in an elastic material stress is proportional to strain, the ratio of stress and strain is constant. This constant is called the modulus (i.e. measure) of elasticity and is usually denoted by E. It is usually expressed in N/mm^2:

$$\frac{\text{Stress}}{\text{Strain}} = E \qquad \frac{W/A}{\delta/l} = E$$

hence $E = \dfrac{Wl}{A\delta}$

where W = direct force,
l = gauge length,
A = original cross-sectional area of member, and
δ = change in length of member.

The modulus of elasticity gives an indication of the 'stiffness' of a material. The greater the value of E, the greater the resistance to deformation (i.e. to lengthening, shortening or bending) of the member: that is, a large stress is required to

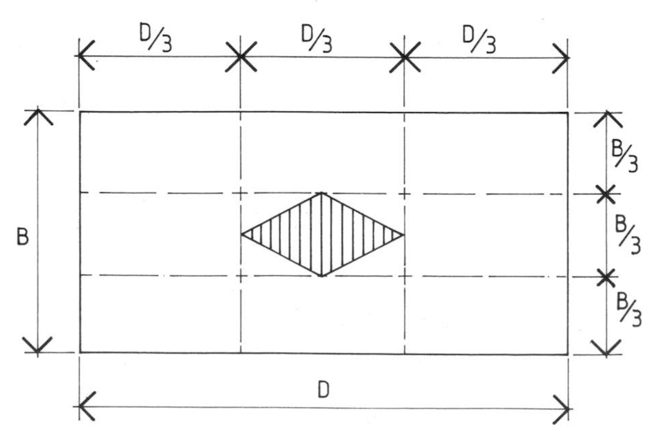

Figure 1.3 The middle-third rule (plan of a pier, etc.)

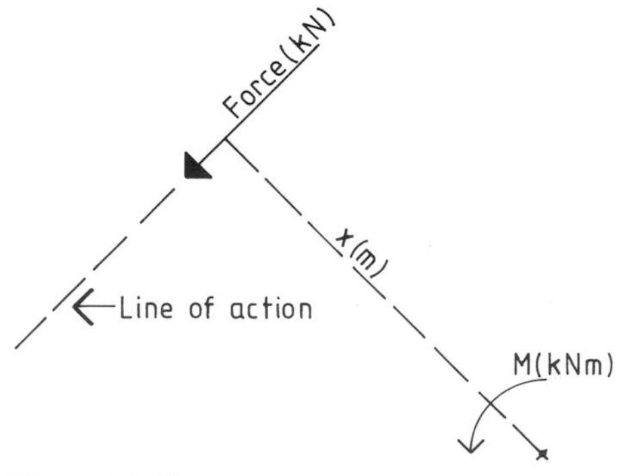

Figure 1.4 The moment of a force

produce a small strain.

The modulus of elasticity must not be confused with 'elastic modulus', which is used in certain steelwork tables in place of 'section modulus' to distinguish between the elastic and plastic section moduli.

Moment of a force

The moment of a force is the turning effect or leverage of that force about a given point or axis. It is measured by multiplying the force by the perpendicular distance of its line of action from the point or axis and is generally expressed in N mm or kN m (Figure 1.4).

Moment bending

When a member or portion of a member is so restrained or fixed that an applied moment cannot cause rotation about the point under consideration, bending is caused and the moment is then called a bending moment. In Figure 1.5 the bending moment at AA is the sum of the moments of the forces acting about that point. In this case $+Vx - wx^2/2$.

Moment of inertia

This is a term used in dynamics for rotating parts such as flywheels and armatures. A flywheel has inertia (i.e. a reluctance to having its state of rest or of motion changed) and a certain moment is required to cause it to rotate. This moment depends on the mass of the material and its arrangement with respect to the axis of rotation. The contribution of each particle of material to the total energy of the wheel depends on the weight of the particle and the square of its distance from the axis of rotation. The moment of inertia is defined as the sum of all these products taking all particles into account. This can be expressed mathematically as

$$I = \sum my^2$$

where m = mass of particles,
y = distance from axis, and
Σ = sum of.

In deriving beam design formulae the expression Σay^2 is arrived at, where a is the area of a small element at a distance y from the axis of bending and, because of its similarity with my^2, it is also called the moment of inertia. However, it might be more accurate to call it the second moment of area.

In Figure 1.6 the second moment of area about XX is

$$I_{XX} = \sum ay^2$$

$$a^1y_1^2 + a^2y_2^2 + a^3y_3^2 + a^4y_4^2 + \text{etc.}$$

For geometrical figures (rectangles, circles, etc.) general formulae for Σay^2 or I can be found by integrating ay^2 through the limits of $+y$(max) and $-y$(max). For symmetrical shapes, see 'Section properties' (page 9).

Moment of resistance

The bending moment of a beam must be resisted by an internal moment set up by the action of the beam fibres. The maximum bending moment M depends on the loads and reactions. The maximum

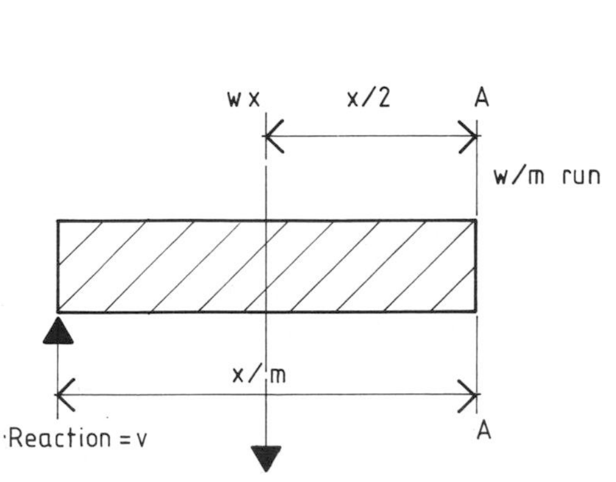

Figure 1.5 A bending moment

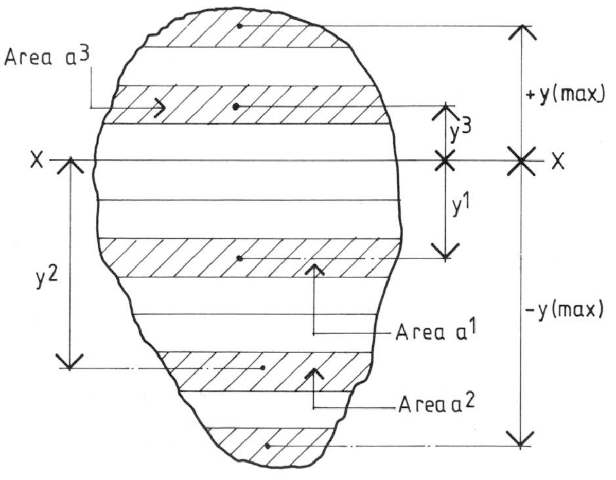

Figure 1.6 The second moment of area

safe moment of resistance M_r a beam can supply depends on the maximum permissible stress σ for the material and the size and shape of the cross section (i.e. the moment of inertia). Normally, for design purposes, the maximum moment of resistance M_r is made equal to the maximum bending moment M. The beam design formula is obtained from the simplified theory of bending:

$$\frac{M}{I} = \frac{\sigma}{y}$$

where M = bending moment or moment of resistance,
I = moment of inertia,
σ = maximum permissible bending stress, and
y = the distance from the neutral axis to the extreme fibres of the beam.

Pinned (hinged) members

A hinge is a form of support that keeps the end of the member in position but allows freedom to rotate. There is therefore no bending moment at a hinge. Depending on the disposition of the loads, the reaction at a hinge can be at any angles (see 'End fixity of beams', page 3).

Radius of gyration

This term was first used in dynamics (e.g. of flywheels). In comparing the effectiveness of such wheels it is imagined that the mass of the wheel is concentrated into one particle at a distance from the centroid of the wheel so that the total moment of inertia is unaltered. The distance of this particle from the centre of the wheel is called the radius of gyration.

With respect to columns which are liable to buckle (bend) the radius of gyration can be thought of as the distance from the axis of bending at which the whole area of cross section can be assumed to be concentrated so that the resistance to bending remains unaltered. The radius of gyration takes into account not only the size of the cross section but also its shape, i.e. the arrangement of the material with respect to the axis of buckling.

Usually, for design of columns, the least radius of gyration is required:

Least radius of gyration

$$= \sqrt{\frac{\text{Least moment of inertia}}{\text{Area of cross section}}}$$

i.e. $r = \sqrt{\frac{I}{A}}$

Roller bearing

When the ends of a member are supported on rollers they are free to move longitudinally and also to rotate. The reaction at a roller bearing is always at right angles to the line of rollers (Figure 1.7).

Section modulus

The moment of resistance M_r of a homogeneous beam depends on the nature of the material and the size and shape of its cross section. The nature of the material is provided for by the permissible

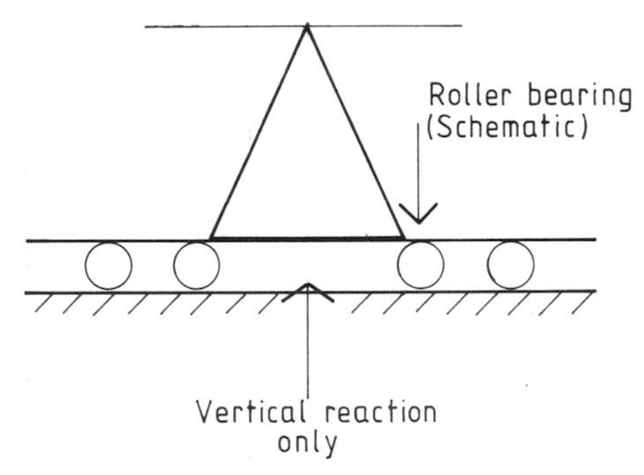

Figure 1.7 Roller bearing

stress σ and the section modulus can be defined as the size–shape factor Z:

$$Z = \frac{I}{y}$$

where I = moment of inertia and
y = distance from neutral axis to extreme fibres of the beam.

Simply (freely) supported beams

See 'End fixity of beams' (page 3).

Slenderness ratio

The slenderness ratio of a steel or timber compression member is usually expressed as

$$\frac{\text{Effective length of the compression member}}{\text{Least radius of gyration}} \text{ i.e. } \frac{l}{r}$$

Tables of permissible stresses for compression members are usually based upon specified end conditions of the compression member, e.g. 'restrained at both ends in position but not in direction' (both ends hinged). In this case the effective length l of the column equals its actual length L. For other conditions of end restraint allowance is made for the different load-carrying capacities of the compression member by using effective lengths as given in BS 5950 or BS 5268. The effective height of a reinforced concrete column may be determined from a table in BS 8110, *The Structural Use of Concrete*, which is based upon the degree of end restraint. In addition, BS 8110 suggests that a more accurate estimate of the effective height may be obtained by considering the actual joint stiffness at each end.

The slenderness ratio of a reinforced concrete column is the effective height divided by the side of the column about both axes. If the ratios about both axes are less than 15 (braced) and 10 (unbraced) the column is considered as short; otherwise it is slender.

Stress

Stress may be thought of as the internal 'distress' of a member as a result of the application of external loads. It is the resistance set up by the particles of the member in opposition to the breaking tendency of the loads. Stress is expressed as load per unit area W/A, e.g. N/mm^2, kN/mm^2 or kN/m^2. In simple bending

$$\frac{M}{I} = \frac{\sigma}{y}$$

and hence

$$\sigma = \frac{My}{I}$$

where M = bending moment or moment of resistance,
I = moment of inertia,
y = distance from neutral axis to the extreme fibre, and
σ = maximum permissible bending stress.

Strain

A member cannot be subjected to a stress without being deformed, i.e. strained. The most important strains are elongation (due to tension) and shortening (due to compression). Such strains are defined as change in length δ divided by original unloaded length L of the member:

$$\text{Strain} = \frac{\delta}{L}$$

Yield stress

This applies particularly to mild steel. If a bar of steel is subjected to a tensile force it behaves 'elastically' until a certain stress (the yield stress) is reached. The bar then stretches a great deal and becomes more or less plastic. A greater load produces increasingly larger extensions until the bar finally breaks.

1.3 Loading

Minimum imposed floor loads

BS 6399: Part 1 provides information on dead and imposed loads. All floor slabs are designed to carry the appropriate distributed or concentrated imposed loads designated in the British Standard. Beams are to be designed to carry a distributed load appropriate to their use. If a value is not given for a concentrated load it may be assumed that the distributed load is adequate for design purposes (Table 1.4).

Reduction in total distributed imposed floor loads

Reductions in imposed loads given in Tables 1.5 and 1.6 may be taken in designing columns, piers, walls and beams as well as their supports and foundations. The reductions given in Table 1.5 do not apply to roofs. Where the floor is designed for 5 kN/m^2 or more the reduction shown in Table 1.6 may be taken providing the loading assumed is not less than it would have been if all floors had been designed for 5 kN/m^2 with no reductions.

Flat roofs and sloping roofs up to 10°

If access, in addition to that needed for cleaning and repair, is to be provided, the imposed load is 1·5 kN/m^2 measured on plan or a 1·8 kN concentrated load, whichever produces the greater stress. If no access is provided other than that required for cleaning and repair, the imposed load is 0·75 kN/m^2 measured on plan or a 0·9 kN concentrated load, whichever produces the greater stress.

Sloping roofs

If the slope is greater than 10° with no access provided other than that needed for cleaning and repair, the following imposed loads apply:

1 Up to 30° a load of 0·75 kN/m^2 measured on plan or a vertical load of 0·9 kN on a square with a 300 mm side.
2 No imposed load to be considered if the slope is greater than 75°. For slopes between 30° and 75° the imposed load is found by linear interpolation.

See BS 6399: Part 1 concerning deflection as the design criterion.

Table 1.4 Minimum imposed floor loads

Type of building or structure	*Distributed load* *kN/m²*	*kgf/m²*	*Concentrated load over a 300 mm square* *kN*	*kgf*
Assembly buildings				
with fixed seating	4·0	408	—	—
without fixed seating	5·0	510	3·6	367
Bedrooms				
domestic buildings	1·5	153	1·4	143
hotels	2·0	204	1·8	184
institutions	1·5	153	1·8	184
Book stores	2·4	245	7·0	714
	for each metre of storage height			
Clubs				
assembly areas with fixed seating	4·0	408	—	—
assembly areas without fixed seating	5·0	510	3·6	367
Colleges				
classrooms	3·0	306	2·7	275
dining rooms	2·0	204	2·7	275
dormitories	1·5	153	1·8	184
gymnasia	5·0	510	3·6	367
Corridors, etc. and footbridges between buildings				
subject to crowd loadings	4·0	408	4·5	459
subject to loads greater than crowds	5·0	510	4·5	459
Dance halls	5·0	510	3·6	367
factories and similar buildings	5·0	510	4·5	459
	7·5 or	765 or	6·7	683
	10·0	1020	9·0	918
Garages				
car parking for vehicles not exceeding 2500 kg	2·5	255	9·0	918
all repair workshops and parking for vehicles exceeding 2500 kg	5·0	510	9·0	918
Hospitals				
wards, utility rooms	2·0	204	1·8	184
operating theatres	2·0	204	4·5	459
Hotels				
bars and vestibules	5·0	510	—	—
kitchens	3·0	306	4·5	459
Houses	1·5	153	1·4	143
Offices				
filing and storage	5·0	510	4·5	459
general offices	2·5	255	2·7	275
offices with computing equipment, etc.	3·5	357	4·5	459
Stairs				
houses less than three stories	1·5	153	1·8	184
all other buildings – the same as the floors to which they give access	3·0 to 5·0	306 to 510		
Stationery stores	4·0	408	9·0	918
	for each metre of storage height			

Table 1.5 Reduction in total distributed imposed floor load

Number of floors including roof carried by member	1	2	3	4	5–10	10+
Percentage reduction in total distributed imposed load on all floors carried by the member under consideration	0	10	20	30	40	50

Table 1.6 Reduction in total distributed imposed floor load

Area supported (m^2)	40	80	120	160	200	240
Percentage reduction in total distributed imposed load	0	5	10	15	20	25

Table 1.7 Horizontal loads on parapets and balustrades (not including Public Assembly class)

Use	*A horizontal UDL (kN/m run)*	*A UDL applied to the infill (kN/m^2)*	*A point load applied to part of the infill (kN)*
Light access stairs, etc. less than 600 mm wide	0·22	N/A	N/A
Stairs, balconies, ramps, landings or floors within, or serving exclusively, one dwelling	0·36	0·5	0·25
Stairs in residential buildings not covered by the above cases	0·36	1·0	0·50

Horizontal loads on parapets and balustrades (not including Public Assembly class)

See Table 1.7.

Wind loading

CP3: Chapter V: Part 2: 1972 provides information on wind loads. The treatment of wind loading is now very much more complicated than in previous codes. It is therefore not possible to summarize the requirements of this code adequately, and reference must be made to the code if such information is required.

1.4 Bending moments, shear force, deflections and shear properties

Mass densities of materials

See Table 1.8.

Maximum bending moments, shear forces and deflections

See Table 1.9.

Section properties

In section properties the following symbols are used:

A = area of cross section (mm^2),

y = distance from neutral axis to extreme fibres (mm). (It is this distance which is also used in the expression $M/I = \sigma/y$.)

XX and YY are axes which pass through the centroid of the section.

I_{XX} and I_{YY} = moments of inertia (i.e. second moments of area) about the axes XX and YY (mm^4)

Z_{XX} and Z_{YY} = section moduli about the axes XX and YY (mm^3)

r_{XX} and r_{YY} = radius of gyration about axes XX and YY (mm)

$$r = \sqrt{\frac{I}{A}}$$

Table 1.8 Mass densities of materials

Material	kg/m^3
Aluminium	2771
Asbestos cement	1922–2082
Asphalt	2082
Bitumen roofing felt	593
Brass	8426
Brickwork, commons	2000
heavy pressed brick	2240
engineering	2400
Cement	1441
Concrete, plain	2300
reinforced	2400
Copper	8730
Cork	128–240
Felt, roofing	593
Fibre building board	160–400
Floors – hollow-clay blocks with concrete ribs between blocks and 40 mm concrete topping	1600
Glass, plate	2787
Lead	11 325
Plaster, acoustic	800
fibrous	430
gypsum	1920
Steel, mild	7849
Stone, limestone	2082–2243
sandstone	2195–2403
granite	2595–2931
Timber, oak	721–961
pitchpine	673
Douglas fir	529

Table 1.9 Maximum bending moments, shearing forces and deflections

Loading	*Maximum bending moment*	*Maximum shearing force*	*Maximum deflection*
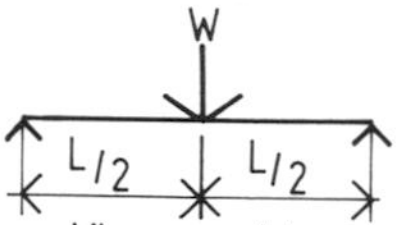 W; L/2, L/2	$\frac{WL}{4}$	$\frac{W}{2}$	$\frac{WL^3}{48EI}$
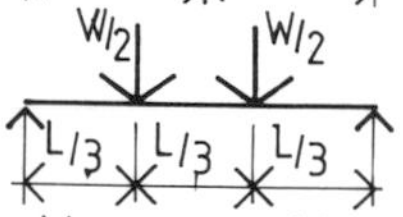 W/2, W/2; L/3, L/3, L/3	$\frac{WL}{6}$	$\frac{W}{2}$	$\frac{23WL^3}{1296EI}$
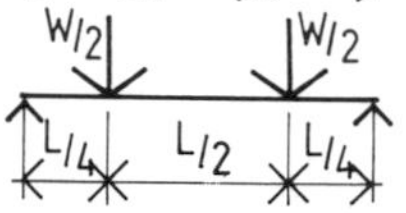 W/2, W/2; L/4, L/2, L/4	$\frac{WL}{8}$	$\frac{W}{2}$	$\frac{11WL^3}{768EI}$
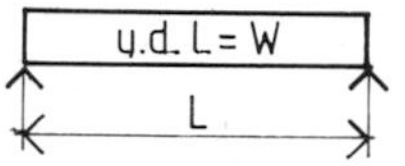 u.d.L = W; L	$\frac{WL}{8}$	$\frac{W}{2}$	$\frac{5WL^3}{384EI}$
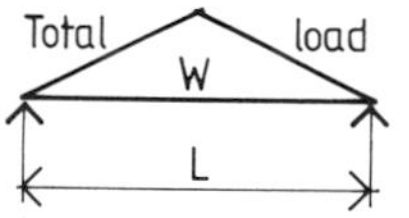 Total load W; L	$\frac{WL}{6}$	$\frac{W}{2}$	$\frac{WL^3}{60EI}$
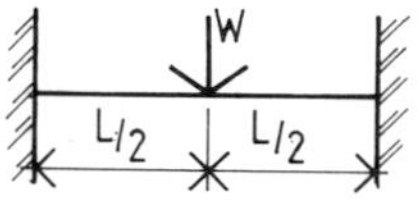 W; L/2, L/2	$\frac{WL}{8}$ (at supports and at midspan)	$\frac{W}{2}$	$\frac{WL^3}{192EI}$
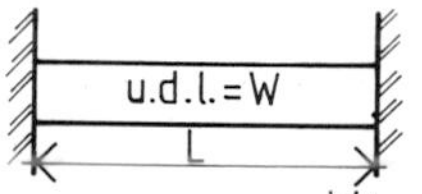 u.d.l. = W; L	$\frac{WL}{12}$ at supports $\frac{WL}{24}$ at midspan	$\frac{W}{2}$	$\frac{WL^3}{384EI}$
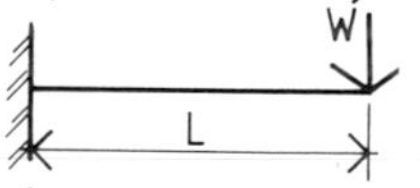 W; L	WL	W	$\frac{WI^3}{3EI}$
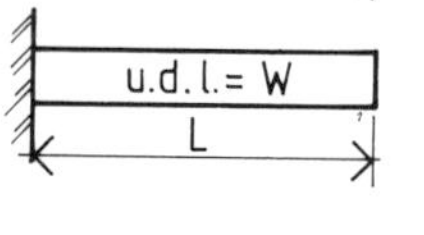 u.d.l. = W; L	$\frac{WL}{2}$	W	$\frac{WL^3}{8EI}$

Properties of a rectangle (Figure 1.8)

$$A = BD$$

$$y_1 = \frac{D}{2} \qquad y_2 = \frac{B}{2}$$

$$I_{XX} = \frac{BD^3}{12} \qquad I_{YY} = \frac{DB^3}{12}$$

$$Z_{XX} = \frac{BD^2}{6} \qquad Z_{YY} = \frac{DB^2}{6}$$

$$r_{XX} = \sqrt{\frac{I_{XX}}{A}} = 0{\cdot}289\sqrt{D}, \; r_{YY} = 0{\cdot}289\sqrt{B}.$$

Properties of a hollow rectangle (Figure 1.9)

$$A = BD - bd$$

$$y_1 = \frac{D}{2} \qquad y_2 = \frac{B}{2}$$

$$I_{XX} = \frac{BD^3 - bd^3}{12} \qquad I_{YY} = \frac{DB^3 - db^3}{12}$$

$$Z_{XX} = \frac{BD^3 - bd^3}{6D} \qquad Z_{YY} = \frac{DB^3 - db^3}{6B}$$

$$r_{XX} = \sqrt{\frac{I_{XX}}{A}} \qquad r_{YY} = \sqrt{\frac{I_{YY}}{A}}$$

Properties of an I-section (Figure 1.10)

$$A = BD - bd$$

$$y_1 = \frac{D}{2} \qquad y_2 = \frac{B}{2}$$

$$I_{XX} = \frac{BD^3 - bd^3}{12} \qquad I_{YY} = \frac{2TB^3}{12} + \frac{dt^3}{12}$$

$$Z_{XX} = I_{XX} \div \frac{D}{2} \qquad Z_{YY} = I_{YY} \div \frac{B}{2}$$

$$r_{XX} = \sqrt{\frac{I_{XX}}{A}} \qquad r_{YY} = \sqrt{\frac{I_{YY}}{A}}$$

Properties of a rectangle with axis on base (Figure 1.11)

$$A = BD$$

$$I_{VV} = \frac{BD^3}{3}$$

Properties of a triangle with axis through centroid (Figure 1.12)

$$A = \frac{BD}{2}$$

$$y = \frac{2D}{3}$$

$$I_{XX} = \frac{BD^3}{36}$$

$$Z_{XX} = \frac{BD^2}{24}$$

$$r_{XX} = \sqrt{\frac{D}{18}} = 0{\cdot}236D$$

Properties of a triangle with axis on base or apex (Figure 1.13)

$$A = \frac{BD}{2}$$

$$I_{VV} = \frac{BD^3}{12} \qquad I_{UU} = \frac{BD^3}{4}$$

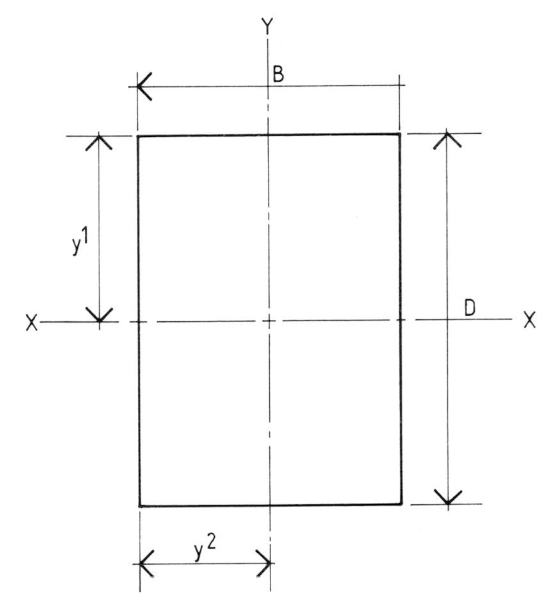

Figure 1.8 Properties of a rectangle

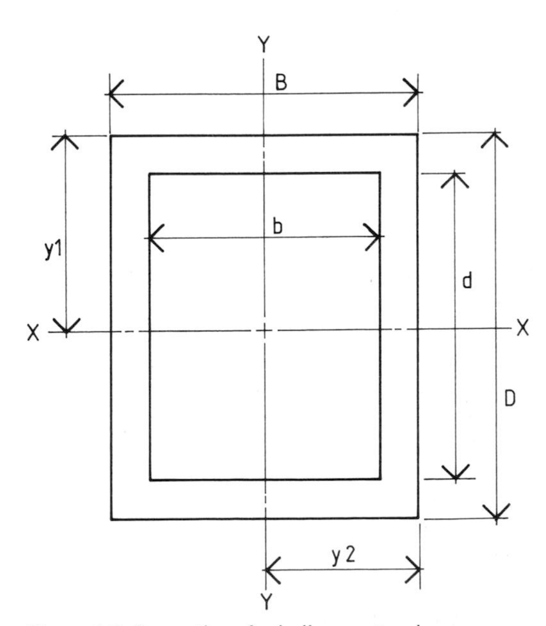

Figure 1.9 Properties of a hollow rectangle

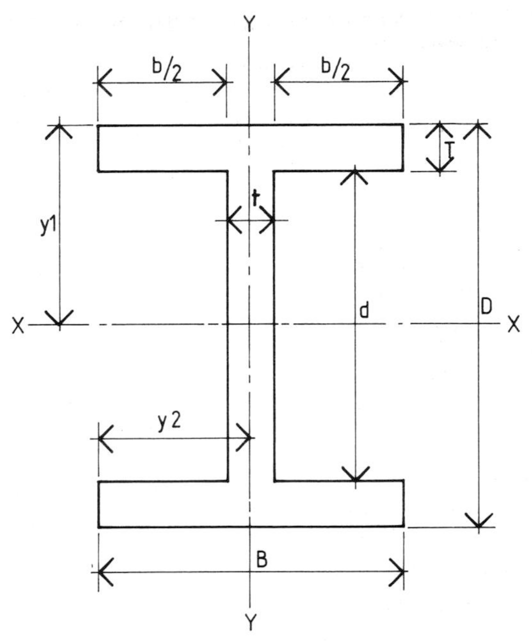

Figure 1.10 Properties of an I-section

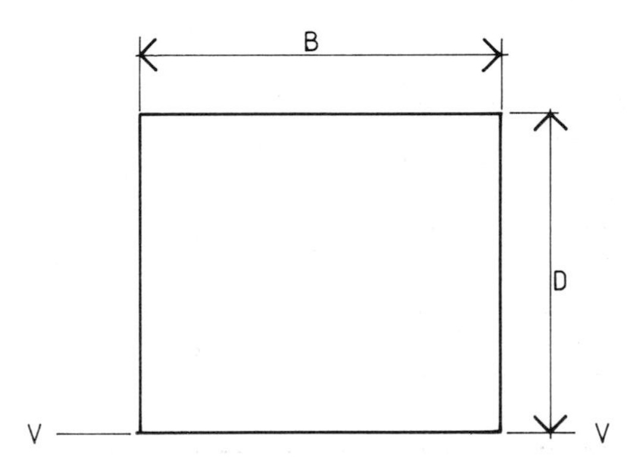

Figure 1.11 Properties of a rectangle with axis on base

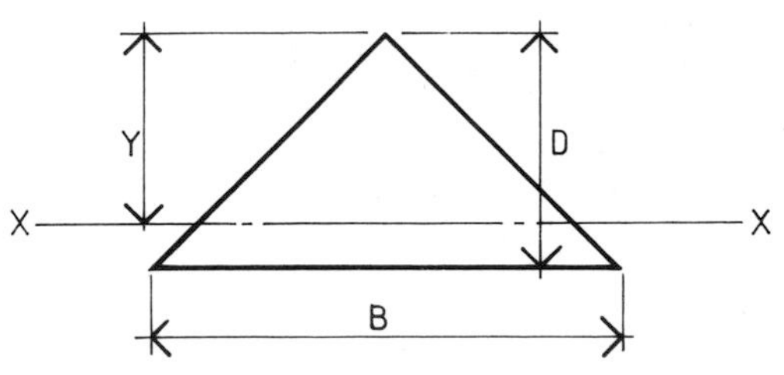

Figure 1.12 Properties of a triangle with axis through centroid

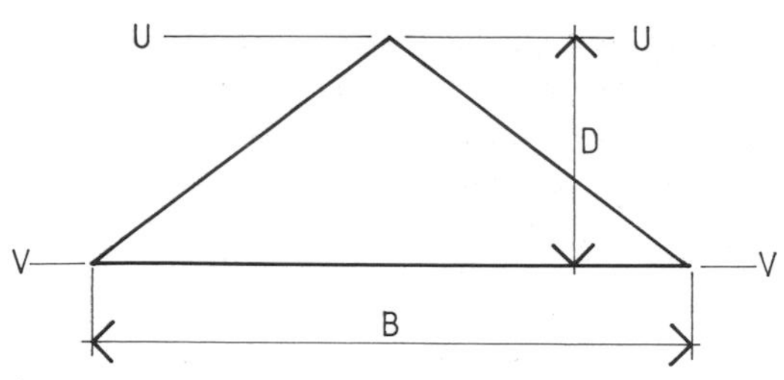

Figure 1.13 Properties of a triangle with axis on base or apex

Properties of a circle with axis through centroid (Figure 1.14)

$$A = \frac{D^2\pi}{4}$$

$$y = \frac{D}{2}$$

$$I_{XX} = I_{YY} = \frac{D^4\pi}{64}$$

$$Z_{XX} = Z_{YY} = \frac{D^3\pi}{32}$$

$$r_{XX} = r_{YY} = \frac{D}{4} = 0{\cdot}25D$$

Properties of a hollow circle (i.e. ring) (Figure 1.15)

$$A = \frac{\pi(D^2 - d^2)}{4}$$

$$y = \frac{D}{2}$$

$$I_{XX} = I_{YY} = \frac{\pi(D^4 - d^4)}{64}$$

$$Z_{XX} = Z_{YY} = \frac{\pi(D^4 - d^4)}{32D}$$

$$r_{XX} = r_{YY} = \sqrt{\frac{(D^2 + d^2)}{4}}$$

$$= 0{\cdot}25\sqrt{(D^2 + d^2)}$$

Properties of a semi-circle with axis through centroid (Figure 1.16)

$$A = \frac{D^2\pi}{8}$$

$$y = R\left(1 - \frac{4}{3\pi}\right)$$

$$I_{XX} = R^4\left(\frac{\pi}{8} - \frac{8}{9\pi}\right) = 0{\cdot}1093R^4$$

$$r_{XX} = \sqrt{\frac{I_{XX}}{A}} = 0{\cdot}2643R$$

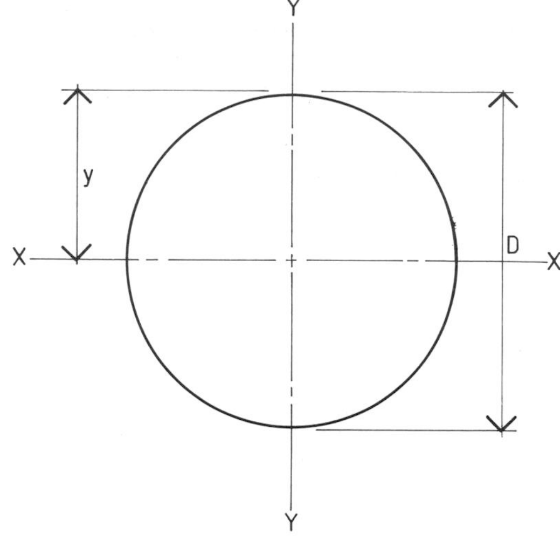

Figure 1.14 Properties of a circle with axis through centroid

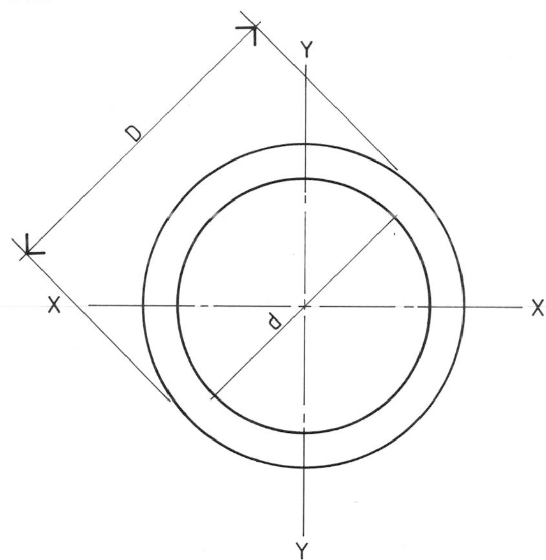

Figure 1.15 Properties of a hollow circle (i.e. ring)

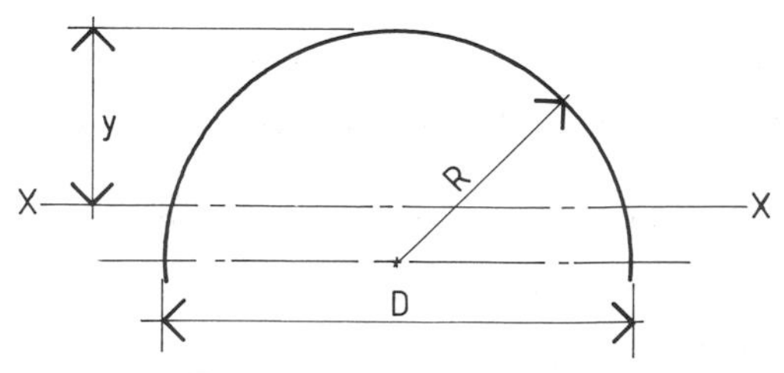

Figure 1.16 Properties of a semicircle with axis through centroid

Moments of inertia of unsymmetrical sections: build-up of rectangles

Usually properties such as section modulus, moment of inertia, etc., are required about axes which pass through the centroid (neutral axis) of the section. When considering an unsymmetrical section such as a flanged beam it is usual to apply the parallel axis theorem to calculate the moment of inertia. The first step, therefore, is to position these axes: the moments of inertia can then be found by using the parallel axis theorem.

Parallel axis theorem

With reference to Figure 1.17,

$$I_{XX} = I_{cg} + Ah^2$$

where I_{XX} = moment of inertia of figure about XX,
I_{cg} = moment of inertia about axis through centroid of section
A = area of section,
h = distance between the parallel axes.

This formula is true for any shape of section. For rectangles

$$I_{XX} = \frac{BD^3}{12} + BDh^2$$

where B is the dimension parallel to the axes cg and XX.

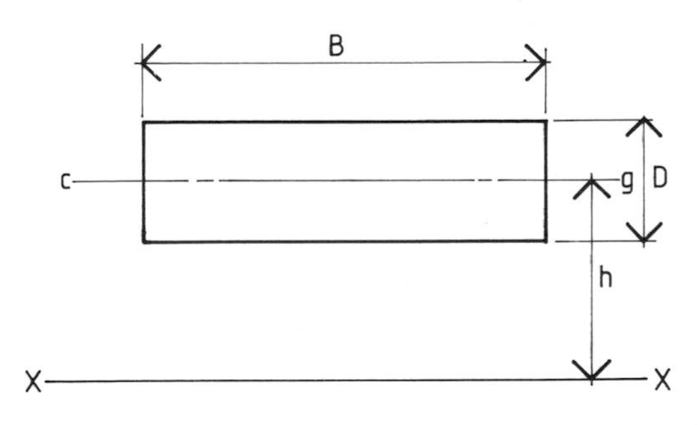

Figure 1.17 Properties of a section remote from an axis

Example 1

Determine I_{XX} for the section shown in Figure 1.18.

Divide the section into two portions. Hence

Area of flange $= A_1$

Area of web $= A_2$

Find the position of axis XX and take moments about AA (see Figure 1.19). Therefore

$$\bar{x} = \frac{\text{Total first moment}}{\text{Total area}} = \frac{A_1x_1 + A_2x_2}{A_1 + A_2}$$

I_{XX} for the section can be found by the parallel axis theorem (see Figure 1.18):

$$I_{XX}\ (\text{flange}) = \frac{BT^3}{12} + BTh_1^2$$

$$I_{XX}\ (\text{web}) = \frac{td^3}{12} + dth_2^2$$

$$\text{Total } I_{XX} = I_{XX}\ (\text{flange}) + I_{XX}\ (\text{web})$$

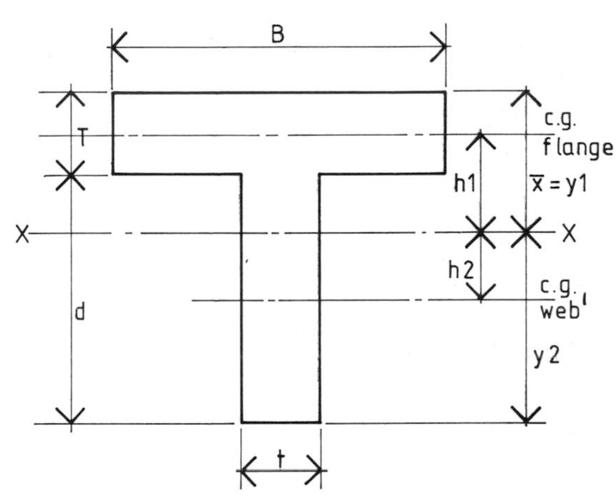

Figure 1.18 Properties of a T-section

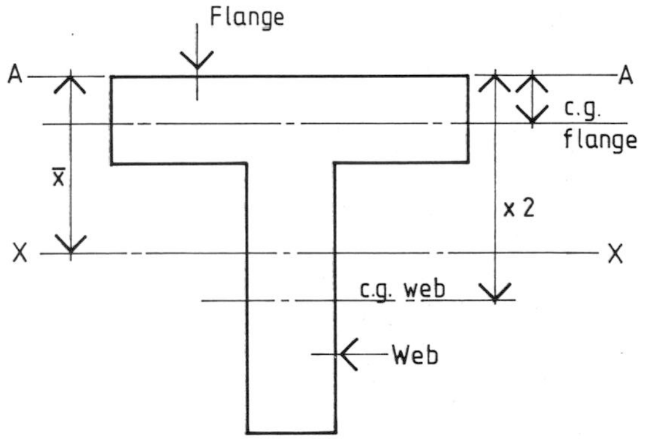

Figure 1.19 Position of XX axis in a T-section

Example 2

Determine I_{XX} for the section in Figure 1.20. Divide the section into three portions, i.e.

Area of top flange $= A_1$

Area of web $= A_2$

Area of bottom flange $= A_3$

Find the position of axis XX and take moments about AA (see Figure 1.21). Therefore

$$\bar{x} = \frac{A_1x_1 + A_2x_2 + A_3x_3}{A_1 + A_2 + A_3}$$

I_{XX} for the section can be found by the parallel axis theorem (see Figure 1.20):

$$I_{XX}\text{ (top flange)} = \frac{B_1T_1^3}{12} + B_1T_1h_1^2$$

$$I_{XX}\text{ (web)} = \frac{td^3}{12} + dth_2^2$$

$$I_{XX}\text{ (bottom flange)} = \frac{B_2T_2^3}{12} + B_2T_2h_3^2$$

$$\text{Total } I_{XX} = I_{XX}\text{ (top flange)} + I_{XX}\text{ (web)} + I_{XX}\text{ (bottom flange)}$$

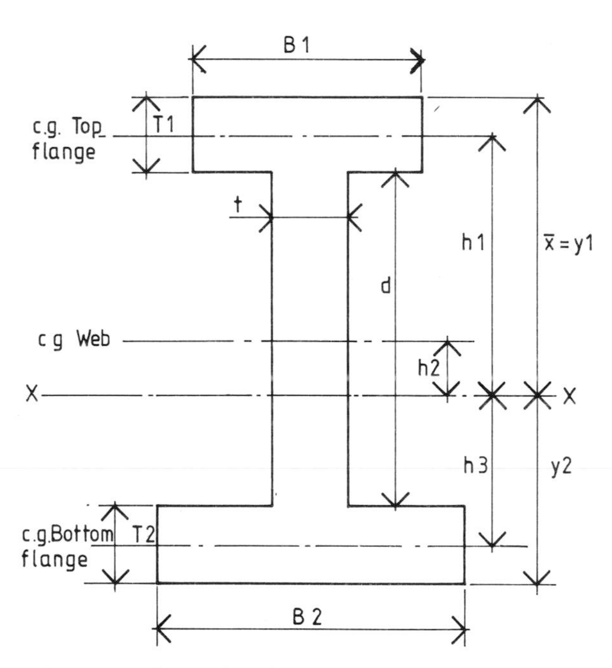

Figure 1.20 Properties of an asymmetric I-section

Section modulus of unsymmetrical sections

When it is necessary to determine the section modulus (Z_{XX}) for the unsymmetrical shape as shown in Figure 1.18 it will be noted that two values are obtained:

1. $Z_{XX} = \dfrac{I_{XX}}{y_1}$ and
2. $Z_{XX} = \dfrac{I_{XX}}{y_2}$

For simple beam design it is usual to use the least value of Z_{XX}, which in this case will be (2).

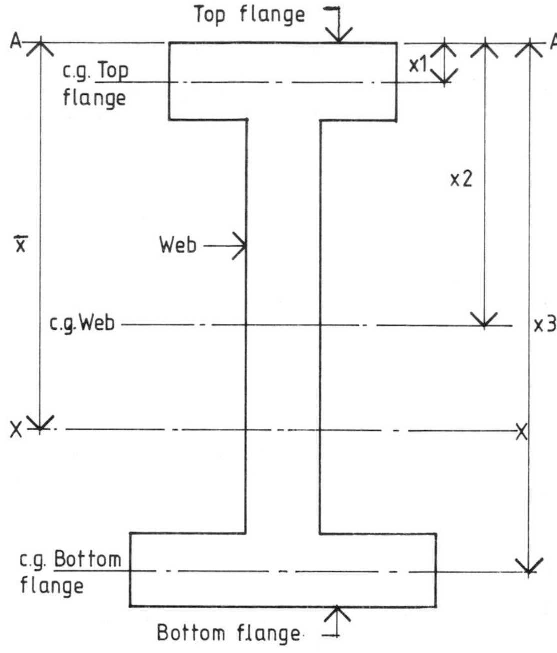

Figure 1.21 Properties of XX axis in an asymmetric I-section

1.5 British standards and codes of practice

A list of useful British Standards and Codes of Practice is given in Table 1.10.

Table 1.10 British Standards and Codes of Practice

Number	*Title*
CP 3: Chapter V: Part 2	Loading – Wind loads
CP 102	Protection of buildings against water from the ground
CP 117: Part 1	Simply supported beams in building
CP 118	The structural use of aluminium
CP 297	Precast concrete cladding (non-load bearing)
CP 2005	Sewerage
BS 648	Schedule of weights of building materials
BS 4978	Timber grades for structural use
BS 5268: Part 1	Limit state design for timber
BS 5268: Part 2	Permissible stress design, materials and workmanship
BS 5268: Part 3	Trussed rafter roofs
BS 5268: Part 4	Fire resistance of timber
BS 5268: Part 5	Preservative treatments for constructional timber
BS 5268: Part 6	Trussed frame wall design
BS 5400: Part 5	Beams for bridges
BS 5337	Code of practice for the structural use of concrete for retaining aqueous liquids
BS 5628: Part 1	Structural use of masonry – Unreinforced masonry
BS 5628: Part 2	Structural use of reinforced and prestressed masonry
BS 5628: Part 3	Materials and components Design and workmanship
BS 6031	Earthworks
BS 6399: Part 1	Loading – Dead and imposed loads
BS 8004	Foundations
BS 8110: Part 1	Structural use of concrete – Design, materials and construction
BS 8110: Part 2	Structural use of concrete – Special circumstances
BS 8110: Part 3	Structural use of concrete – Design Charts
BS 8301	Building drainage

2

Structural steelwork

This chapter deals with the structural design of the following structural members:

1. Beams subject to bending only;
   Case (a) Fully restrained compression flange
   Case (b) Partially restrained compression flange
   Case (c) Unrestrained compression flange
   Case (d) The conservative method for unrestrained or partially unrestrained compression flange
   Case (e) Cased member.
2. Columns subject to axial load only
3. Discontinuous single- and double-angle struts subject to axial load only
4. Cased column subject to axial load only
5. Columns in a simple multi-storey construction
6. Simple tension members
7. Simple connections
8. Column base plate with an axial compressive load.

2.1 General

The above design cases follow the recommendations for the design of structural steel elements in simple construction as described in section 4 of Part 1 of BS 5950; *Structural Use of Steelwork in Building*. When finally completed the code will consist of nine parts: however, only Parts 1 (1985), 2 (1985), 4 (1985) and 5 (1987) had been issued by the time this book was published. The full list of Parts 1 to 9 is as follows:

Part 1: Code of practice for design in simple and continuous construction: hot rolled sections.
Part 2: Specification for materials, fabrication and erection: hot rolled sections.
Part 3: Code of practice for design in composite construction.
Part 4: Code of practice for design of floors with profiled steel sheeting.
Part 5: Code of practice for design in cold formed sections.
Part 6: Code of practice for design in light gauge sheeting, decking and cladding.
Part 7: Specification for materials and workmanship: cold formed sections.
Part 8: Code of practice for design of fire protection for structural steelwork.
Part 9: Code of practice for stressed skin design.

Definitions

Listed below are a selection of definitions relative to the work covered by this chapter. Symbols are defined when used in the text.

Beam. A member predominantly subject to bending.

Buckling resistance. Limit of force or moment which a member can withstand without buckling.
Cantilever. A beam which is fixed at one end and free to deflect at the other.
Capacity. Limit of force or moment which may be applied without causing failure due to yielding or rupture.
Column. A vertical member of a structure carrying axial load and possibly moments.
Compact cross section. A cross section which can develop the plastic moment capacity of the section but in which local buckling prevents rotation at constant moment.
Compound section. Constructed by interconnecting one or more sections or plates and sections to form a single member.
Dead load. All loads of constant magnitude and position that act permanently, including self-weight.
Design strength. The yield strength of the material multiplied by the appropriate partial factor.
Edge distance. Distance from the centre of a fastener hole to the nearest edge of an element.
Effective length. Length between points of effective restraint of a member multiplied by a factor to take account of the end conditions and loading.
End distance. Distance from the centre of a fastener hole to the edge of an element parallel to the direction in which the fastener bears.
Factored load. Specified load multiplied by the relevant partial factor.
Friction grip connection. A bolted connection which relies on friction to transmit shear between components.
Imposed load. Load on a structure or member, other than wind load, produced by the external environment and intended occupancy or use.
Lateral restraint. For a beam: restraint which prevents lateral movement of the compression flange. For a compression member: restraint which prevents lateral movement of the member in a particular plane.
Longitudinal. Along the length of the member.
Pitch. Distance between centres of fasteners lying in the direction of stress.
Plastic cross section. A cross section which can develop a plastic hinge with sufficient rotation capacity to allow redistribution of bending moments within the structure.
Semi-compact cross section. A cross section in which the stress in the extreme fibres should be limited to yield because local buckling would prevent development of the plastic moment capacity in the section.
Serviceability limit states. Those limit states which, when exceeded, can lead to the structure being unfit for its intended use.
Slender cross section. A cross section in which yield of the extreme fibres cannot be attained because of premature local buckling.
Slenderness. The effective length divided by the radius of gyration.
Slip resistance. Limit of shear that can be applied before slip occurs in a friction grip connection.
Strength. Resistance to failure by yielding or buckling.
Strut. A member of a structure carrying predominantly compressive axial load.
Transverse. Direction perpendicular to the stronger of the rectangular axes of the member.
Ultimate limit state. That state which, if exceeded, can cause collapse of part or whole of the structure.

2.2 Beams subject to bending only

For this section the following assumptions are made:

1. The method of construction is 'simple'.
2. The beam is not subjected to a destabilizing load.
3. The design loading is composed of dead and imposed loading only.
4. The beam sections class is plastic or compact with equal flanges.

The structural design is presented in the following stages:

1. The design loading and resulting ultimate bending moment and shearing forces.
2. Obtaining a basic section size and its classification.
3. Establishing the available lateral and torsional restraint for the beam.
4. Design cases;
   (a) Full lateral and torsional restraint;
   (b) Partial lateral and torsional restraint;
   (c) Unrestrained.
5. Local buckling.
6. Deflection.
7. Cased section.

The above stages as applied to a particular case are illustrated in Figure 2.1.

Stage 1

1. Establish the span L of the beam. This should be taken as the distance between the effective points of support (Figure 2.2).
2. Calculate the characteristic dead and imposed loading the beam is to support (see 'Loading' in Chapter 1) for appropriate values of dead and imposed loading. If the beam section is not known at this stage make an allowance of, say, 5% of the dead load for the beam self-weight.
3. Calculate the design loading as follows:

$\omega' = \gamma_{f_i}\omega_i + \gamma_{f_d}\omega_d$ (kN/m)

$\omega' = 1{\cdot}6\omega_i + 1{\cdot}4\omega_d$

$\gamma_{f_{i,d}}$: load factors from Table 2, BS 5950

4. Calculate the maximum bending moment due to the design loading. (See Table 1.9, Chapter 1, for typical simple beam load cases and maximum bending moment values.) The maximum bending moment is the 'design moment' m (in units of kN m).
5. Calculate the maximum shearing force F_v (again, see Table 1.9 for typical cases). The design shear force F_v is in kN units. As a member must be checked for:

   (a) Design moment and co-existent shear, and
   (b) Design shear force and co-existant moment,

it is also necessary to calculate the value of the shear force, if it exists, at the position of the design moment M.

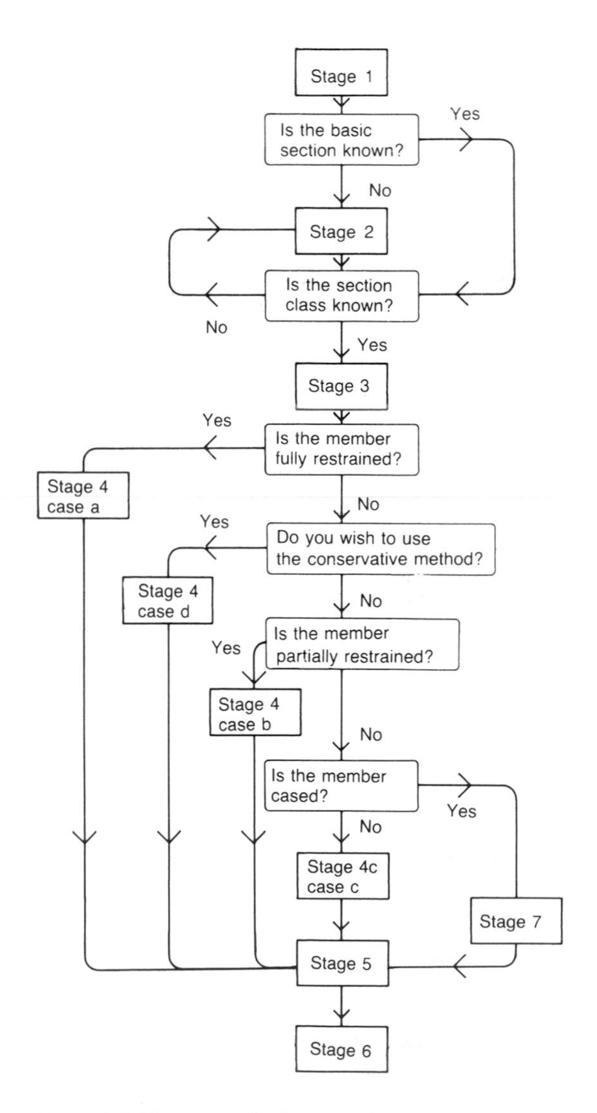

Figure 2.1 Flow chart for beam design

Stage 2

The basic section

Essentially, the process of the design of a structural steel beam is one of checking a basic section for adequacy in the appropriate limit states. The basic section may be obtained using one of the following methods, depending on the restraint condition of the beam. (See Stage 3 for guidance on restraint conditions.)

Compression flange fully restrained laterally

In this case the following condition must be met:

$M < M_c$

where M = the design moment, and
M_c = the moment capacity of the section.

Now, assuming the shear load is low:

$M_c = p_y S$

and

$S_{reqd} = \dfrac{M_c}{p_y}$ (S_{reqd} is in cm^3 units).

From the section tables choose a section whose plastic modulus S about the major axis XX is such that

$S_X \geqslant S_{reqd}$

This will now be the basic section.

Compression flange not fully restrained laterally

In this case the lateral torsional buckling of the section before the full moment capacity is reached may occur. Here the following condition must be satisfied:

$\bar{M} = mM$

where $\bar{M}$ is an equivalent uniform moment and m a factor applied to M to obtain $\bar{M}$, and

$\bar{M} \leqslant M_b$

where M_b is the buckling resistance moment of the section. Also

$M_b = S_x p_b$

where p_b is the bending strength of the section. p_b is now a reduced value of the design strength, p_y which, at this stage of the calculations, can only be estimated for the purpose of obtaining a basic section. If a value of, say, $1{\cdot}25p_b$ is assumed for p_y, and for m, the equivalent uniform moment factor, a value of 1·0 then

$$M = S_X\, 0{\cdot}8p_y$$

and

$$S_{Xreqd} = \frac{M}{0{\cdot}8p_y}$$

Using the section tables, choose a basic section whose plastic moduli S_X is such that

$$S_X \geqslant S_{Xreqd}$$

Section classification

The classification of structural steel sections is based on the ratio of width to thickness of the section parts as described in Table 7 and Figure 3 in BS 5950. However, as noted at the beginning of this section, it is assumed that the beam is formed from a section whose class is plastic or compact. This assumption will be satisfied if the section is chosen from the range of Universal Beam sections and joists in the section properties tables at the end of this chapter.

Stage 3

Lateral restraint

Lateral restraint to the compression flange may be provided along its complete length by the supported floor provided that there is a positive connection between the floor and the beam's compression flange. Figure 2.3 shows typical forms of construction and their appropriate level of restaint to the beam's compression flange.

Lateral restraint may also be provided at points along the beam's length by individual members supported by the beam as long as the connection is positive with the compression flange.

Torsional restraint

The typical end connections shown in Figure 2.4 illustrate the forms of positive connections required to produce torsional restraint. As with lateral restraint, torsional restraint may also occur at positions along the length of the beam by incoming beams.

Stage 4

Case (a) Fully restrained member (Figure 2.5(a))

1. Calculate the shear capacity P_V for the section

   $$P_V = 0{\cdot}6p_yA_V$$

   where p_y = the design strength from Table 2.1,
   A_V = the shear area (= tD),
   t = the thickness of the web, and
   D = the depth of the section.

Table 2.1 Design strength p_y

BS 4360 Grade	*Thickness less than or equal to (mm)*	*Design strength p_y (N/mm²)*
43	16	275
A, B and C	40	265
	100	245

2. Check that $F_V \leqslant 0{\cdot}6P_V$. If $F_V > 0{\cdot}6P_V$ see clause 4.2.6 of the code.
3. For the case of $F_V \leqslant 0{\cdot}6P_V$ the moment capacity of the section is given by

   $$M_c = p_yS_X \text{ but } \leqslant 1{\cdot}2p_yZ_X$$

   where S is the plastic modulus about the axis of bending.
4. Check that $M \leqslant M_c$ (i.e. the ultimate applied moment) is less than or equal to the moment capacity of the section.

Case (b) Partially restrained member (Figure 2.5(b))

In this form of construction, where between the points of restraint the beam is possibly subject to lateral torsional buckling, the nature of the loading pattern will dictate the approach to be adopted to obtain the equivalent uniform moment and slenderness correction factors m and n. The load patterns shown in Figure 2.5(b) are:

Case b(i) Member not loaded in the length between adjacent points of restraint; i.e. the beam is supporting its self-weight only between the points of restraint.

Case b(ii) Member loaded in the length between adjacent points of restraint. This condition is further subdivided into:

b(ii)1 The loading on the beam between the adjacent points of restraint is substantially concentrated within the middle fifth of the unrestrained length; and

b(ii)2 The loading is not concentrated within the middle fifth of the unrestrained length.

Calculate the equivalent uniform moment $\bar{M}$ and buckling resistance moment M_b. The following condition is to be satisfied in all three cases:

$$\bar{M} = mM \leqslant M_b$$

where $\bar{M}$ = the equivalent uniform moment,
m = the equivalent uniform moment factor,
M = the maximum applied moment in the unrestrained length,
M_b = the buckling resistance moment of the section,

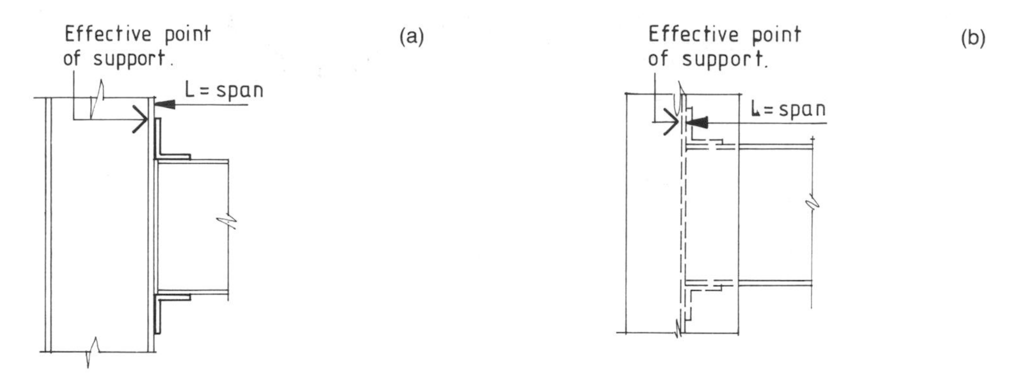

Figure 2.2 Beam connected to (a) column flange and (b) column web

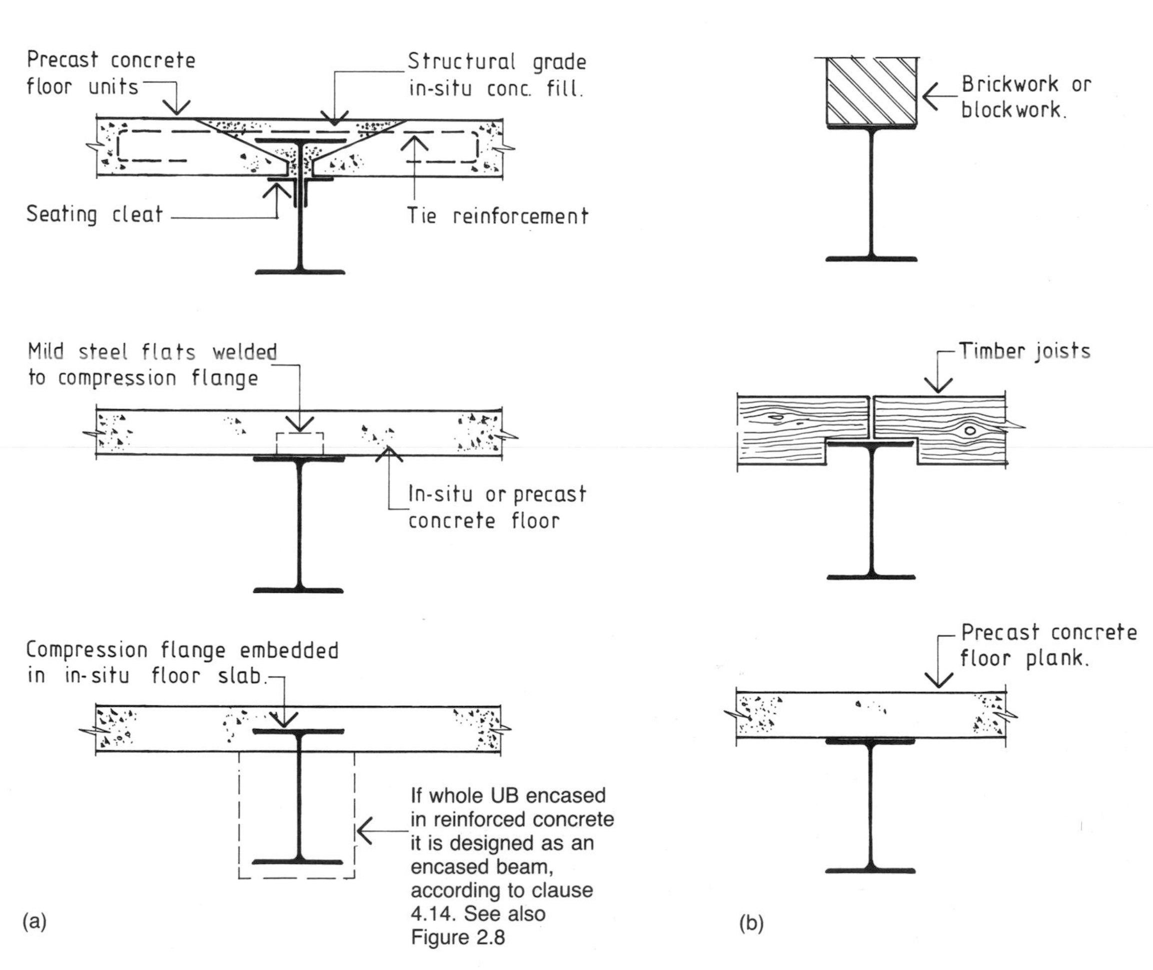

Figure 2.3 Typical examples of (a) fully restrained and (b) non-restrained compression flanges

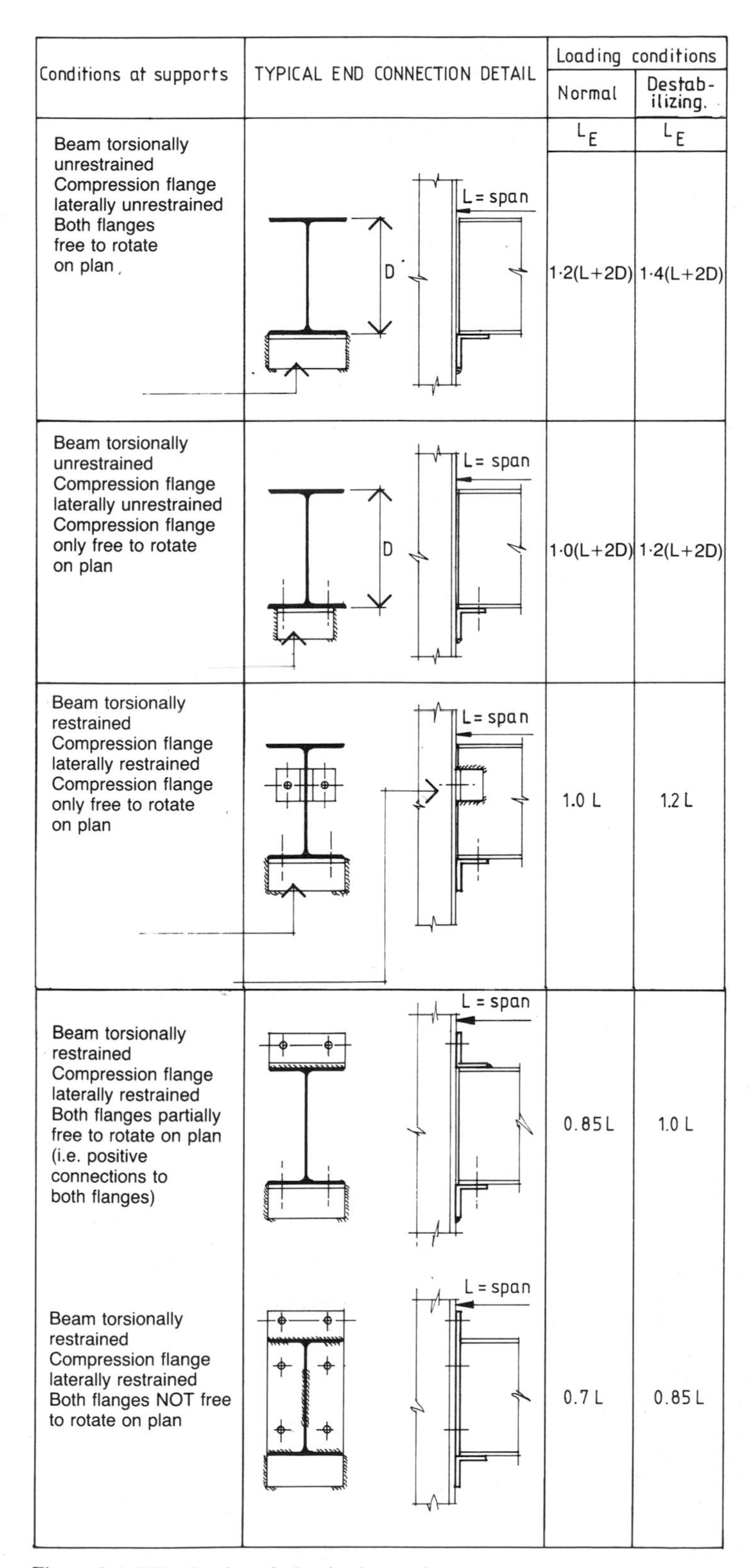

Conditions at supports	TYPICAL END CONNECTION DETAIL	Loading conditions: Normal L_E	Loading conditions: Destabilizing L_E
Beam torsionally unrestrained Compression flange laterally unrestrained Both flanges free to rotate on plan	L = span; D	1·2(L+2D)	1·4(L+2D)
Beam torsionally unrestrained Compression flange laterally unrestrained Compression flange only free to rotate on plan	L = span; D	1·0(L+2D)	1·2(L+2D)
Beam torsionally restrained Compression flange laterally restrained Compression flange only free to rotate on plan	L = span	1.0 L	1.2 L
Beam torsionally restrained Compression flange laterally restrained Both flanges partially free to rotate on plan (i.e. positive connections to both flanges)	L = span	0.85 L	1.0 L
Beam torsionally restrained Compression flange laterally restrained Both flanges NOT free to rotate on plan	L = span	0.7 L	0.85 L

Figure 2.4 Effective length L_E for beams between supports

and

$M_b = S_X p_b$

where S_X = the plastic modulus about the axis of bending, and

p_b = the bending strength based on the yield stress (p_y) and the equivalent slenderness (λ_{LT})

Case b(i)

1. Reading Table 2.2, m is to be obtained from Table 2.3 and n, the slenderness correction factor, is equal to 1·0. Each length of beam between adjacent points of restraint should now be checked to establish the critical length; i.e. the largest value of $\bar{M}$:

Length L_{A-B}

$$\beta = \frac{M_A}{M_B} = \frac{0}{M_B} = 0$$

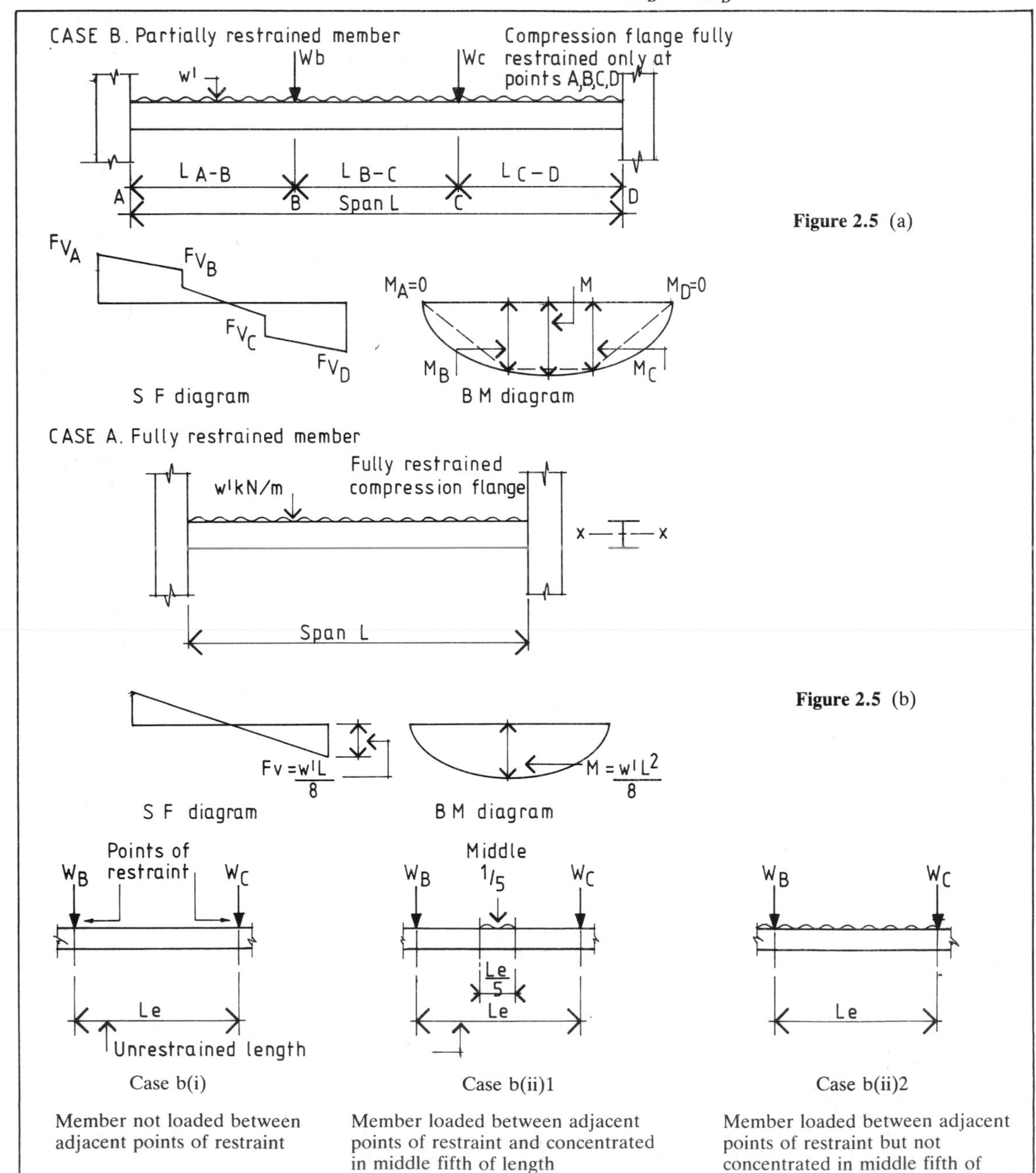

Figure 2.5 (a)

Figure 2.5 (b)

Case b(i)

Member not loaded between adjacent points of restraint

Case b(ii)1

Member loaded between adjacent points of restraint and concentrated in middle fifth of length

Case b(ii)2

Member loaded between adjacent points of restraint but not concentrated in middle fifth of length

Table 2.2 Equivalent moment and slenderness correction factors

Description (loading condition between adjacent points of restraint)	*Case (see Figure 2.5(b))*	*Equivalent moment factor m*	*Slenderness factor n*
Member not loaded between adjacent points of restraint	b(i)	Table 2.3	1·0
Member loaded between adjacent points of restraint and concentrated in middle fifth of length	b(ii)1	1·0	Table 2.4(a)
Member loaded between adjacent points of restraint but not concentrated in middle fifth of length	b(ii)2	1·0	Table 2.4(b)

Note: For cantilevers, $m = 1{\cdot}0$ and $n = 1{\cdot}0$.

Table 2.3a Slenderness factor υ for symmetrical flange sections (from Table 14, BS 5950)

λ/x	0·5	1·0	1·5	2·0	2·5	3·0	3·5	4·0	4·5	5·0	5·5	6·0	6·5	7·0	7·5
υ	1·00	0·99	0·97	0·96	0·93	0·91	0·89	0·86	0·84	0·82	0·79	0·77	0·75	0·73	0·72

λ/x	8·0	8·5	9·0	9·5	10	11	12	13	14	15	16	17	18	19	20
υ	0·70	0·68	0·67	0·65	0·64	0·61	0·59	0·57	0·55	0·53	0·52	0·50	0·49	0·48	0·47

λ: the minor axis slenderness ($= l_e/\upsilon_y$).
L_e is the effective length of the section between adjacent points of restraint.
r_Y is the radius of gyration about the minor axis.
x: the torsional index for the section obtained from the section properties table at the end of the chapter.

Table 2.3b Equivalent uniform moment factor, *m*

	β	m
Beta positive	1·0	1·00
	0·9	0·95
	0·8	0·90
	0·7	0·85
	0·6	0·80
	0·5	0·76
	0·4	0·72
	0·3	0·68
	0·2	0·64
	0·1	0·60
Beta negative	0·0	0·57
	−0·1	0·54
	−0·2	0·51
	−0·3	0·48
	−0·4	0·45
	−0·5 to −1·0	0·43

Note 1. The values of m in this table apply only to end moments applied to beams of *uniform* section with *equal* flanges. In all other cases $m = 1{\cdot}0$.
Note 2. Values of m for intermediate values of β may be interpolated, or determined from the equation: $m = 0{\cdot}57 + 0{\cdot}33\beta + 0{\cdot}10\beta^2$: but not less than 0·43.

From Table 2.3b read off the value of m for the above β value:

Maximum moment on length $L_{A-B} = M_{A-B}$
and

$\bar{M}_{A-B} = mM_{A-B}$
Length L_{B-C}

$$\beta = \frac{M_B}{M_C} \text{ or } \frac{M_C}{M_B} \text{ for } \beta \leqslant 1{\cdot}0$$

Read off the m value for β from Table 2.3b.

Maximum moment on length $L_{B-C} = M_{B-C}$

and

$\bar{M}_{B-C} = mM_{B-C}$

Length L_{C-D}

$$\beta = \frac{M_C}{M_D} = \frac{M_C}{0} = 0$$

Read off the m value for β from Table 2.3b.

Maximum moment on length $L_{C-D} = M_{C-D}$

and

$\bar{M}_{C-D} = mM_{C-D}$

2. It is now necessary to check that

   $\bar{M} \leqslant M_b$

   for each length of unrestrained beam. M_b, the buckling resistance moment for each unrestrained length, is calculated as follows:

   $M_b = S_X p_b$

   where S_X is the plastic modulus of the section about the axis of bending and p_b is the bending strength from Table 2.5 based on p_y and λ_{LT} values.

   Equivalent slenderness of span λ_{LT}

   $\lambda_{LT} = nuv\lambda$

   where n = slenderness correction factor (= 1·0, see Table 2.2 for load case i),

   u = buckling parameter for the section (this value is obtained from the section properties tables at the end of the chapter),

   v = a slenderness factor based on λ/x and N, where $N = 0{\cdot}5$ (see Table 2.3a for the appropriate value of v), and

   λ = the minor axis slenderness as used for v in Table 2.3.

   Having obtained values of n, u, v and λ, calculate the equivalent slenderness of the span, i.e.:

   $\lambda_{LT} = nuv\lambda$

3. Bending strength p_b. The value of p_b is obtained from Table 2.5 using the calculated value of λ_{LT} and the appropriate design strength of the steel p_y from Table 2.1.
4. Buckling resistance moment M_b. Using the appropriate value of p_b for each length of beam between adjacent points of restraint, calculate the buckling resistance moment M_b and check against the equivalent uniform moment $\bar{M}$. From the example in Figure 2.5(a) the following conditions must be satisfied:

   $M_{b_{A-B}} = S_X p_{b_{A-B}} \geqslant \bar{M}_{A-B}$

   $M_{b_{B-C}} = S_X p_{b_{A-B}} \geqslant \bar{M}_{B-C}$

   $M_{b_{C-D}} = S_X p_{b_{C-D}} \geqslant \bar{M}_{C-D}$

   Check that the condition

   $M \leqslant M_c$

Table 2.4 Slenderness correction factor, *n*
(a) Members with applied loading substantially concentrated in the middle fifth of the unrestrained span

$\gamma = M/M_0$	*β positive* 1·0	0·8	0·6	0·4	0·2	0·0
+50·00	1·00	0·96	0·92	0·87	0·82	0·77
+10·00	0·99	0·99	0·94	0·90	0·85	0·80
+5·00	0·98	0·98	0·97	0·93	0·89	0·84
+2·00	0·96	0·95	0·95	0·95	0·94	0·94
+1·50	0·95	0·95	0·94	0·94	0·93	0·93
+1·00	0·93	0·92	0·92	0·92	0·92	0·91
+0·50	0·90	0·90	0·90	0·89	0·89	0·89
0·00	0·86	0·86	0·86	0·86	0·86	0·86

(b) Members with applied loading other than as for (a)

$\gamma = M/M_0$	*β positive* 1·0	0·8	0·6	0·4	0·2	0·0
+50·00	1·00	0·96	0·92	0·87	0·83	0·77
+10·00	0·99	0·98	0·95	0·91	0·86	0·81
+5·00	0·99	0·98	0·97	0·94	0·90	0·85
+2·00	0·98	0·98	0·97	0·96	0·94	0·92
+1·50	0·97	0·97	0·97	0·96	0·95	0·93
+1·00	0·97	0·97	0·97	0·96	0·96	0·95
+0·50	0·96	0·96	0·96	0·96	0·96	0·95
0·00	0·94	0·94	0·94	0·94	0·94	0·94

Note: The above tables are extracts from Tables 15 and 16 of the code showing n for positive values of β and γ only.

is satisfied at the point of:

(a) The maximum bending moment and co-existent shear on the whole span, and
(b) The maximum shear force and co-existent moment on the whole span,

as described in case (a), page 18).

Case b(ii)1: (Figure 2.5(b)), Calculation of bending strength p_b for M_b

The loading is concentrated in the middle fifth of the span between adjacent points of restraint.

1. Reading Table 2.2, n, the slendernes correction factor, is obtained from Table 2.4(a) and m, the equivalent uniform moment factor, is 1·0.
2. Table 2.4(a) is used to obtain n when the loading between the adjacent points of restraint is substantially concentrated on the middle fifth of the span:

   (a) $\beta = \dfrac{M_B}{M_C}$ such that $\beta \leq 1{\cdot}0$.

   (b) $\gamma = \dfrac{M}{M_0}$

   where M is the maximum value of the bending moments at the two points of restraint and M_0 is the midspan moment on a simply supported span equal to the unrestrained length and the load supported by that span only.

   Using the β and γ values, read off the value for the slenderness correction factor n.
3. Equivalent slenderness of span

   $\lambda_{LT} = nuv\lambda$

   Using the value of n obtained from Table 2.4(a) and u, v and λ as calculated for case b(i), obtain λ_{LT}.
4. The bending strength p_b and the buckling resistance moment M_b for the span are now obtained as shown for case b(i), step 4.

Case b(ii)2: Calculation of bending strength p_b for M_b

The beam is loaded between adjacent points of restraint.

1. Reading Table 2.2, n, the slenderness correction factor, is obtained from Table 2.4(b) and m, the equivalent uniform moment factor, is 1·0.
2. The procedure now follows that of case b(ii)1.

Table 2.5 Bending strength p_b(N/mm²)

λ_{LT} \ p_y	245	265	275	325	340	355	415	430	450
30	245	265	275	325	340	355	408	421	438
35	245	265	273	316	328	341	390	402	418
40	238	254	262	302	313	325	371	382	397
45	227	242	250	287	298	309	350	361	374
50	217	231	238	272	282	292	329	338	350
55	206	219	226	257	266	274	307	315	325
60	195	207	213	241	249	257	285	292	300
65	185	196	201	225	232	239	263	269	276
70	174	184	188	210	216	222	242	247	253
75	164	172	176	195	200	205	223	226	231
80	154	161	165	181	186	190	204	208	212
85	144	151	154	168	172	175	188	190	194
90	135	141	144	156	159	162	173	175	178
95	126	131	134	144	147	150	159	161	163
100	118	123	125	134	137	139	147	148	150
105	111	115	117	125	127	129	136	137	139
110	104	107	109	116	118	120	126	127	128
115	97	101	102	108	110	111	117	118	119
120	91	94	96	101	103	104	108	109	111
125	86	89	90	95	96	97	101	102	103
130	81	83	84	89	90	91	94	95	96
135	76	78	79	83	84	85	88	89	90
140	72	74	75	78	79	80	83	84	84
145	68	70	71	74	75	75	78	79	79
150	64	66	67	70	70	71	73	74	75
155	61	62	63	66	66	67	69	70	70
160	58	59	60	62	63	63	65	66	66
165	55	56	57	59	60	60	62	62	63
170	52	53	54	56	56	57	59	59	59
175	50	51	51	53	54	54	56	56	56
180	47	48	49	51	51	51	53	53	53
185	45	46	46	48	49	49	50	50	51
190	43	44	44	46	46	47	48	48	48
195	41	42	42	44	44	44	46	46	46
200	39	40	40	42	42	42	43	44	44
210	36	37	37	38	39	39	40	40	40
220	33	34	34	35	35	36	36	37	37
230	31	31	31	32	33	33	33	34	34
240	29	29	29	30	30	30	31	31	31
250	27	27	27	28	28	28	29	29	29

Case 4c: Unrestrained member (Figure 2.6)

1. Moment capacity check, i.e.

   $M_{max} \leq M_c$

   (a) Maximum moment and co-existent shear at midspan:

   Shear capacity $P_V = 0{\cdot}6p_yA_V$ (p_y; see Table 2.1, $A_V = tD$)

   and as F_V at $M_{max} = 0$, $F_V < 0{\cdot}6P_V$

$M_c = p_y S_X \leq 1{\cdot}2 p_y Z$

check $M_{max} \leq M_C$.

(b) Maximum shear and co-existent moment at support:

Shear capacity P_V

As $M_A = 0$, check $F_{V_A} \leq P_V$.

2. Buckling resistance moment check, i.e.

$\bar{M} = mM_{max} \leq M_b$

(a) Equivalent moment and slenderness correction factors m and n (Table 2.2). From load condition 3:

$m = 1{\cdot}0 \therefore \bar{M} = M_{max}$

Obtaining n from Table 2.4(b):

$$\beta = \frac{M_A}{M_{max}}, \; M_A = 0 \therefore \beta = 0$$

$$\gamma = \frac{M_A}{M_0} \; M_{max} = M_0 \therefore \gamma = 0$$

From Table 2.4(b) $n = 0{\cdot}94$.

(b) Slenderness factor v (Table 2.3). Calculate λ/x, where $\lambda = L_e/r_y$, and from Figure 2.4 $L_e = 1{\cdot}0(L + 2D)$ and x is the torsional index for the beam section. Read off v for the λ/x value from Table 2.3.

(c) From the section properties table obtain the buckling parameter u for the beam section.

(d) Calculate the equivalent slenderness λ_{LT}:

$\lambda_{LT} = nuv\lambda$

(e) Bending strength p_b. From Table 2.5 read off p_b for the above values of λ_{LT} and p_y.

(f) Buckling resistance moment M_b:

$M_b = p_b S_X$

(g) Check that $\bar{M} \leq M_b$.

Case (d) The conservative method for calculating M_b

Clause 4.3.7.7 of the code offers a simplified method of calculating the buckling resistance moment M_b of an unrestrained span between points of restraint. The method is only applicable to equal-flanged rolled sections.

1. The equivalent uniform moment,

$\bar{M} = mM$

where $m = 1{\cdot}0$ and M is the maximum moment between the lateral restraint. Hence $\bar{M} = M$.

2. The buckling resistance moment,

$M_b = P_b S_X$

where S_X is the plastic modulus about the XX axis and P_b the bending strength from Table 2.6 based on λ and x.

3. To obtain P_b two variables need to be calculated:

(a) λ the slenderness of the section

$$\lambda = \frac{l_{e.n}}{r_Y}$$

where l_e is the effective length of the span between the adjacent points of restraint (see Figure 2.4), n is the slenderness correction factor (this value may be taken as being equal to 1·0 or from Table 2.7, according to the loading on the span) and r_Y is the radius of gyration about the minor axis YY, obtained from the section properties table.

(b) x, the torsional index, obtained from the section properties table.

The bending strength P_b is now read off from Table 2.6 using the calculated values of λ and x.

Now calculate $M_b = P_b S_X$ to check that

$\bar{M} = M \leq M_b$

Stage 5: Buckling and bearing capacity of the web

At any position on a beam where a load is transferred through the line of the web the capacity of the web at that position to resist buckling or load crushing at its base must be checked. The two checks are:

1. Web buckling, and
2. Web bearing.

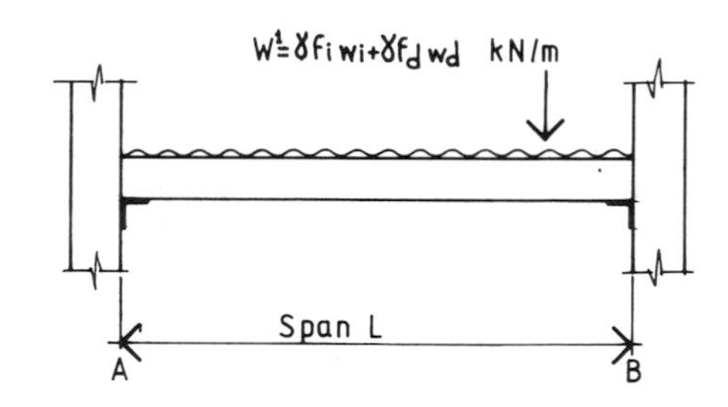

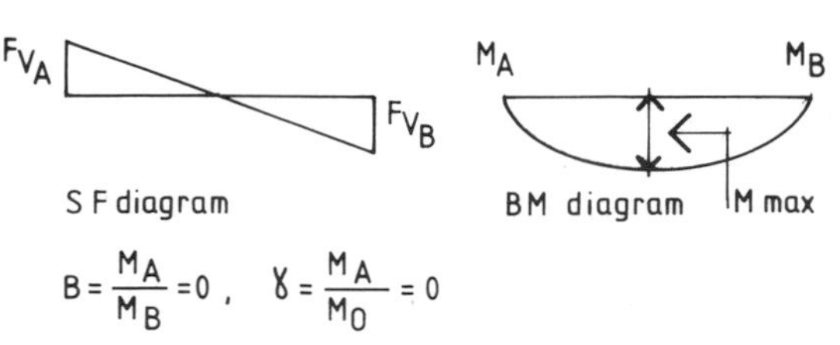

Figure 2.6 Case 4c: unrestrained member

Web buckling (Figure 2.7)

1. Calculate the stiff bearing length b_1.
2. Calculate the dispersion length through the beam section n_1.
3. Note t, the web thickness.
4. Obtain from Table 2.8(c) the compression strength p_c. Table 2.8(c) requires the slenderness λ of the unstiffened web and the design strength of the steel p_y.
   λ: If the flange through which the load or reaction is applied is effectively restrained against;
   (a) Rotation relative to the web,
   (b) Lateral movement relative to the other flange:

   $\lambda = 2{\cdot}5d/t$; if not, $\lambda = 3{\cdot}46d/t$

   p_y: Obtain the value of the design strength of the steel from Table 2.1. Using the above values of λ and p_y, read off from Table 2.8(c) the appropriate value of p_c, the compressive strength.
5. Calculate the web buckling resistance

   $$P_w = (b_1 + n_1)tp_c$$

6. Check that

   $$F_V \leqslant P_w$$

   where F_V is the applied load or reaction.
7. If $F_V > P_w$ then the web will require stiffening. For details of the design of load-carrying stiffeners see clause 4.5.4 of the code.

Web bearing (Figure 2.7)

1. Calculate the stiff bearing length b_1.
2. Calculate the dispersion length through the beam section n_2.
3. Note t, the web thickness.
4. Obtain the design strength of the web p_{y_w}. Take $p_{y_w} = p_y$, the design strength of the steel (Table 2.1).
5. Calculate the local capacity of the web at its connection to the flange:

   $$P_{crip} = (b_1 + n_2)tp_{y_w}$$

6. Check that

   $$F_V \leqslant P_{crip}$$

   where F_V is the applied load or reaction.
7. If $F_V > P_{crip}$ then the web will require stiffening. For details of the design of bearing stiffeners see clause 4.5.4 of the code.

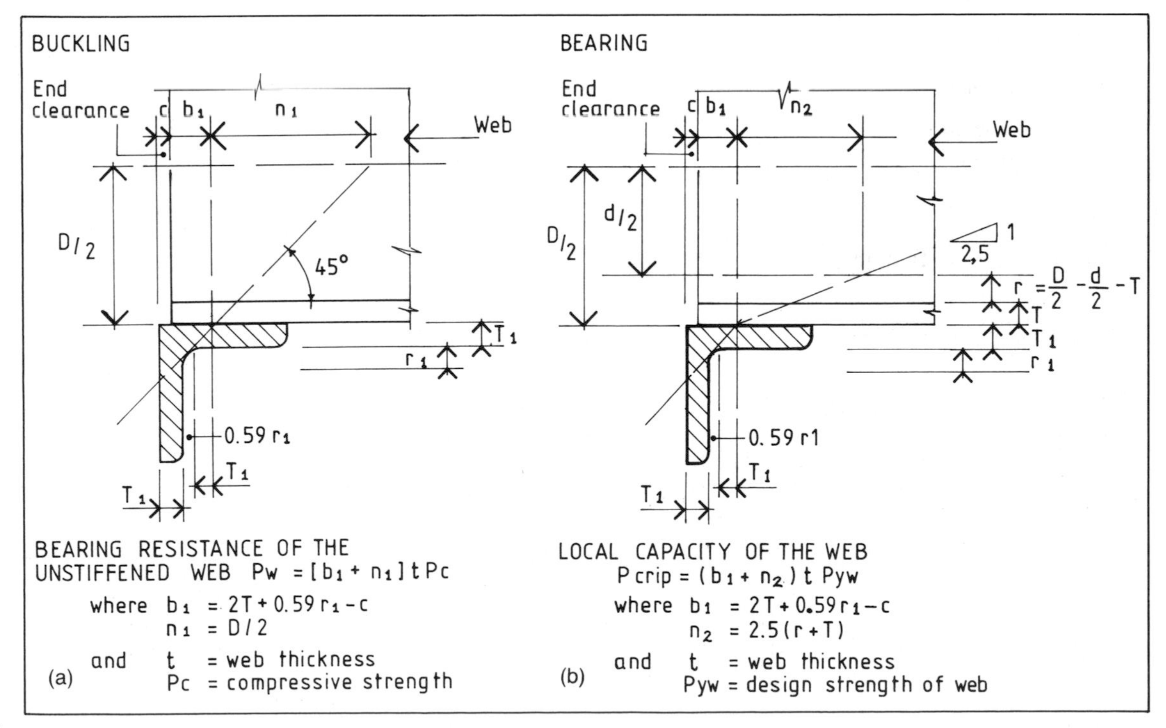

Figure 2.7 Buckling (a) and bearing (b) resistance of an unstiffened web

Table 2.6 Bending strength p_b (N/mm^2) for rolled sections with equal flanges
(a) $p_y = 265\ N/mm^2$

λ \ x	5	10	15	20	25	30	35	40	45	50
30	265	265	265	265	265	265	265	265	265	265
35	265	265	265	265	265	265	265	265	265	265
40	265	265	265	265	265	264	264	264	263	263
45	265	265	261	258	256	255	254	254	254	254
50	265	261	253	249	247	246	245	244	244	244
55	265	255	246	241	238	236	235	235	234	234
60	265	250	239	233	229	227	226	225	224	224
65	265	245	232	225	221	218	216	215	214	214
70	265	240	225	217	212	209	207	205	204	204
75	263	235	219	210	204	200	198	196	195	194
80	260	230	213	202	196	191	189	187	185	184
85	257	226	207	195	188	183	180	178	176	175
90	254	222	201	188	180	175	171	169	167	166
95	252	217	196	182	173	167	163	160	158	157
100	249	213	190	176	166	160	156	153	150	149
105	247	209	185	170	160	153	148	145	143	141
110	244	206	180	164	154	147	142	138	136	134
115	242	202	176	159	148	140	135	132	129	127
120	240	198	171	154	142	135	129	125	123	121
125	237	195	167	149	137	129	124	120	117	115
130	235	191	163	144	132	124	119	114	111	109
135	233	188	159	140	128	119	114	109	106	104
140	231	185	155	124	136	115	109	105	102	99
145	229	182	152	132	120	111	105	101	97	95
150	227	179	148	129	116	107	101	97	93	91
155	225	176	145	125	112	103	97	93	89	87
160	223	173	142	122	109	100	94	89	86	83
165	221	170	139	119	106	97	91	86	83	80
170	219	167	136	116	103	94	88	83	80	77
175	217	165	133	113	100	91	85	80	77	74
180	215	162	130	110	97	88	82	77	74	71
185	213	160	128	108	95	86	79	75	71	69
190	211	157	125	105	92	83	77	73	69	66
195	209	155	123	103	90	81	75	70	67	64
200	207	153	120	101	88	79	73	68	65	62
210	204	148	116	96	84	75	69	64	61	58
220	200	144	112	93	80	71	65	61	58	55
230	197	140	108	89	77	68	62	58	54	52
240	194	136	104	86	74	65	59	55	52	49
250	190	132	101	83	71	63	57	52	49	47

(b) p_y = 275 N/mm²

λ \ x	5	10	15	20	25	30	35	40	45	50
30	275	275	275	275	275	275	275	275	275	275
35	275	275	275	275	275	275	275	275	275	275
40	275	275	275	275	274	273	272	272	272	272
45	275	275	269	266	264	263	263	262	262	262
50	275	269	261	257	255	253	253	252	252	251
55	275	263	254	248	246	244	243	242	241	241
60	275	258	246	240	236	234	233	232	231	230
65	275	252	239	232	227	224	223	221	221	220
70	274	247	232	223	218	215	213	211	210	209
75	271	242	225	215	209	206	203	201	200	199
80	268	237	219	208	201	196	193	191	190	189
85	265	233	213	200	193	188	184	182	180	179
90	262	228	207	193	185	179	175	173	171	169
95	260	224	201	186	177	171	167	164	162	160
100	257	219	195	180	170	164	159	156	153	152
105	254	215	190	174	163	156	151	148	146	144
110	252	211	185	168	157	150	144	141	138	136
115	250	207	180	162	151	143	138	134	131	129
120	247	204	175	157	145	137	132	128	125	123
125	245	200	171	152	140	132	126	122	119	116
130	242	196	167	147	135	126	120	116	113	111
135	240	193	162	143	130	121	115	111	108	106
140	238	190	159	139	126	117	111	106	103	101
145	236	186	155	135	122	113	106	102	99	96
150	233	183	151	131	118	109	102	98	95	92
155	231	180	148	127	114	105	99	94	91	88
160	229	177	144	124	111	101	95	90	87	84
165	227	174	141	121	107	98	92	87	84	81
170	225	171	138	118	104	95	89	84	81	78
175	223	169	135	115	101	92	86	81	78	75
180	221	166	133	112	99	89	83	78	75	72
185	219	163	130	109	96	87	80	76	72	70
190	217	161	127	107	93	84	78	73	70	67
195	215	158	125	104	91	82	76	71	68	65
200	213	156	122	102	89	80	74	69	65	63
210	209	151	118	98	85	76	70	65	62	59
220	206	147	114	94	81	72	66	62	58	55
230	202	143	110	90	78	69	63	58	55	52
240	199	139	106	87	74	66	60	56	52	50
250	195	135	103	84	72	63	57	53	50	47

Stage 6: deflection

1. Calculate the serviceability load to be applied:

   $\omega = \gamma_{f\_i\_}, \gamma_{f\_i\_} = 1{\cdot}0$

   ∴ unfactored imposed load used.
2. Using Table 2.9 (page 34), identify the limiting deflection δ_{lim}.
3. Calculate the actual deflection of the beam, δ. (See Table 1.9 for typical load patterns and maximum deflection.) The modulus of elasticity for the steel should be taken as $E = 205\ \text{kN/mm}^2$ and the moment of inertia I for the section from the section properties table.
4. Check that the actual deflection is less than the limiting deflection:

   $\delta < \delta_{lim}$

Stage 7: cased beam

Figure 2.8 shows the general requirements for an I-section with equal flanges encased in concrete as stated in clause 4.14.1 of the code. According to the restraint conditions, the beam should be checked for the moment capacity and buckling resistance moment, as described in Stage 4, for an uncased section. However, the following variations from Stage 4 should be noted

1. Radius of gyration r_Y.

   For the cased section r_Y should be taken as the greater of:

   (a) $0{\cdot}2\ (B + 100)$ or

   (b) The minimum radius of gyration of the uncased section.

Table 2.7 Slenderness correction factor n for the simplified method

Beam and loads	*Actual bending moment*	n
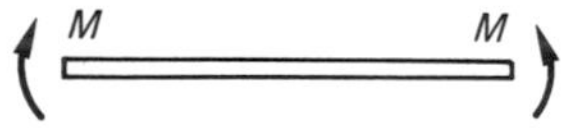 M M	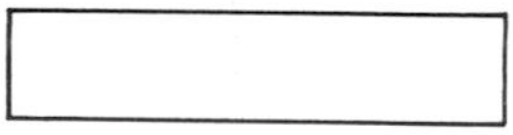	1·0
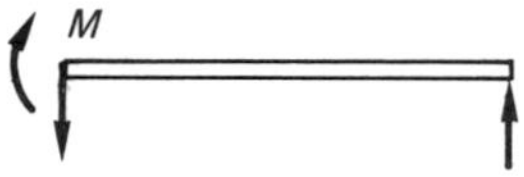 M		0·77
 M M	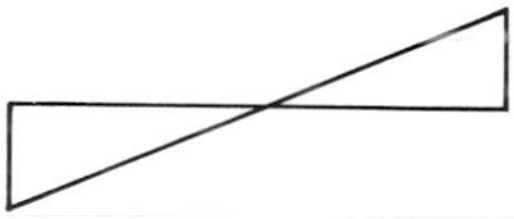	0·65
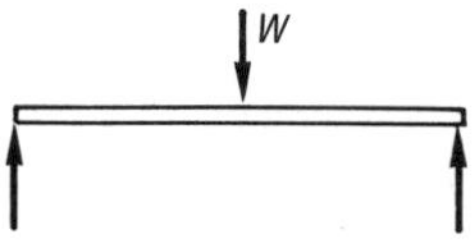 W		0·86
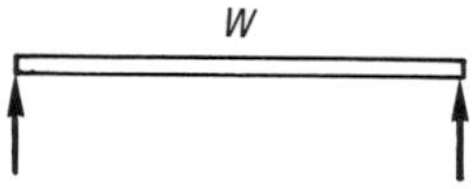 W	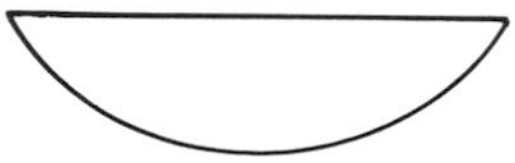	0·94
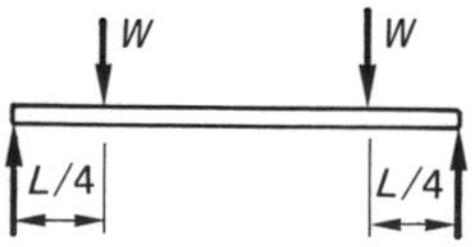 W W L/4 L/4	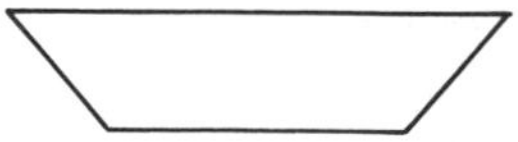	0·94
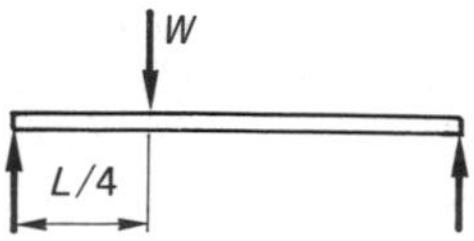 W L/4		0·94

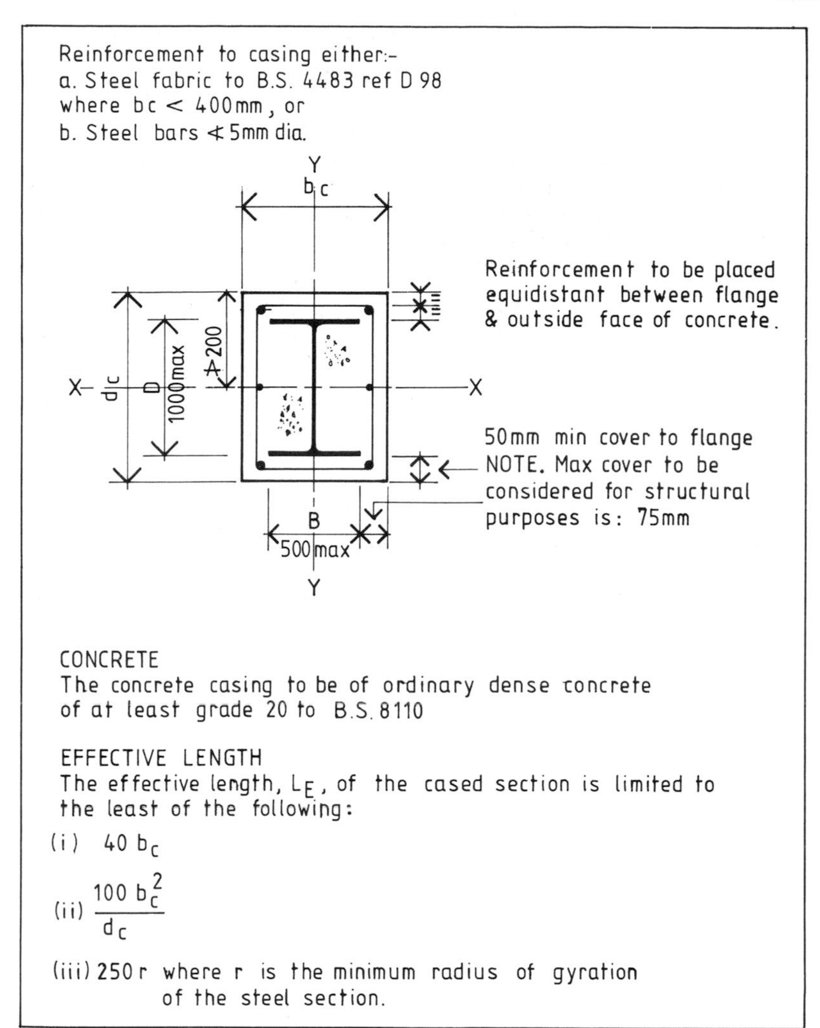

Figure 2.8 General requirements for cased sections

2. The buckling resistance moment of the cased section should not exceed one and a half times that of the uncased one.
3. Deflection. When calculating the actual deflection the effective moment of inertia of the cased section is calculated as follows:

$$I_{cs} = I_s + \frac{I_c - I_s}{\alpha_e}$$

where I_{cs} = second moment of area of the cased section
I_c = second moment of area of the gross concrete section,
I_s = second moment of area of the steel section and,
α_e = the modular ratio.

2.3 Columns subject to axial load only

For this section the following assumptions are made:

1. The design loading is composed of dead and imposed loading only.
2. The column is formed from a Universal Column, Universal Beam or joist rolled sections.
3. For the purpose of calculating the effective length a nominal value based on standard conditions at restraint is assumed. For other conditions of restraint see Appendices D and E of the code.
4. Grade 43 steel is used.

Table 2.8 Compressive strength p_c for struts (N/mm^2)

(a)

λ \ p_y	225	245	255	265	275
15	225	245	255	265	275
20	225	244	254	264	273
25	222	241	251	261	270
30	220	239	248	258	267
35	217	236	245	254	264
40	214	233	242	251	260
42	213	231	240	249	258
44	212	230	239	248	257
46	210	228	237	246	255
48	209	227	236	244	253
50	208	225	234	242	251
52	206	223	232	241	249
54	205	222	230	238	247
56	203	220	228	236	244
58	201	218	226	234	242
60	200	216	224	232	239
62	198	214	221	229	236
64	196	211	219	226	234
66	194	209	216	223	230
68	192	206	213	220	227
70	189	204	210	217	224
72	187	201	207	214	220
74	184	198	204	210	216
76	182	194	200	206	212
78	179	191	197	202	208
80	176	188	193	198	203
82	173	184	189	194	199
84	170	181	185	190	194
86	167	177	181	186	190
88	164	173	177	181	185
90	161	169	173	177	180
92	158	166	169	173	176
94	154	162	165	168	171
96	151	158	161	164	166
98	147	154	157	159	162
100	144	150	153	155	157
102	141	146	149	151	153
104	137	142	145	147	149
106	134	139	141	143	145
108	131	135	137	139	141
110	127	132	133	135	137
112	124	128	130	131	133
114	121	125	126	128	129
116	118	121	123	124	125
118	115	118	120	121	122
120	112	115	116	118	119
122	109	112	113	114	115
124	107	109	110	111	112
126	104	106	107	108	109
128	101	104	105	105	106

(b)

λ \ p_y	225	245	255	265	275
15	225	245	255	265	275
20	224	243	253	263	272
25	220	239	248	258	267
30	216	234	243	253	262
35	211	229	238	247	256
40	207	224	233	241	250
42	205	222	231	239	248
44	203	220	228	237	245
46	201	218	226	234	242
48	199	215	223	231	239
50	197	213	221	229	237
52	195	210	218	226	234
54	192	208	215	223	230
56	190	205	213	220	227
58	188	202	210	217	224
60	185	200	207	214	221
62	183	197	204	210	217
64	180	194	200	207	213
66	178	191	197	203	210
68	175	188	194	200	206
70	172	185	190	196	202
72	169	181	187	193	198
74	167	178	183	189	194
76	164	175	180	185	190
78	161	171	176	181	186
80	158	168	172	177	181
82	155	164	169	173	177
84	152	161	165	169	173
86	149	157	161	165	169
88	146	154	158	161	165
90	143	150	154	157	161
92	139	147	150	153	156
94	136	143	147	150	152
96	133	140	143	146	148
98	130	137	139	142	145
100	127	133	136	138	141
102	124	130	132	135	137
104	122	127	129	131	133
106	119	124	126	128	130
108	116	121	123	125	126
110	113	118	120	121	123
112	111	115	117	118	120
114	108	112	114	115	117
116	105	109	111	112	114
118	103	106	108	109	111
120	100	104	105	107	108
122	98	101	103	104	105
124	96	99	100	101	102
126	94	96	97	99	100
128	91	94	95	96	97

(c)

λ \ p_y	225	245	255	265	275
15	225	245	255	265	275
20	224	242	252	261	271
25	217	235	245	254	263
30	211	228	237	246	255
35	204	221	230	238	247
40	198	214	222	230	238
42	195	211	219	227	235
44	193	208	216	224	231
46	190	205	213	220	228
48	187	202	209	217	224
50	184	199	206	213	220
52	181	196	203	210	217
54	179	193	199	206	213
56	176	189	196	202	209
58	173	186	192	199	205
60	170	183	189	195	201
62	167	179	185	191	197
64	164	176	182	188	193
66	161	173	178	184	189
68	158	169	175	180	185
70	155	166	171	176	181
72	152	163	168	172	177
74	149	159	164	169	173
76	146	156	160	165	169
78	143	152	157	161	165
80	140	149	153	157	161
82	137	146	150	154	157
84	134	142	146	150	154
86	132	139	143	146	150
88	129	136	139	143	146
90	126	133	136	139	142
92	123	130	133	136	139
94	120	127	130	133	135
96	118	124	127	129	132
98	115	121	123	126	129
100	112	118	120	123	125
102	110	115	118	120	122
104	107	112	115	117	119
106	105	110	112	114	116
108	102	107	109	111	113
110	100	104	106	108	110
112	98	102	104	106	107
114	96	99	101	103	105
116	93	97	99	101	102
118	91	95	96	98	100
120	89	93	94	96	97
122	87	91	92	93	95
124	85	88	90	91	92
126	83	86	88	89	90
128	82	84	86	87	88

Table 2.8 (*cont.*)

(a)

λ \ p_y	225	245	255	265	275
130	99	101	102	103	103
135	93	95	96	96	97
140	87	89	90	90	91
145	82	84	84	85	85
150	78	79	79	80	80
155	73	74	75	75	75
160	69	70	70	71	71
165	65	66	67	67	67
170	62	63	63	63	64
175	59	59	60	60	60
180	56	56	57	57	57
185	53	54	54	54	54
190	51	51	51	51	52
195	48	49	49	49	49
200	46	46	47	47	47

(b)

λ \ p_y	225	245	255	265	275
130	89	92	93	94	95
135	84	86	87	88	89
140	79	81	82	83	84
145	75	77	78	78	79
150	71	72	73	74	74
155	67	69	69	70	70
160	64	65	66	66	66
165	60	61	62	63	63
170	57	58	59	59	60
175	55	56	56	56	57
180	52	53	53	54	54
185	49	50	51	51	51
190	47	48	48	48	49
195	45	46	46	46	47
200	43	44	44	44	44

(c)

λ \ p_y	225	245	255	265	275
130	80	82	84	85	86
135	75	78	79	80	81
140	71	74	75	76	76
145	68	70	70	71	72
150	64	66	67	68	68
155	61	63	63	64	65
160	58	59	60	61	61
165	55	56	57	58	58
170	52	54	54	55	55
175	50	51	52	52	53
180	48	49	49	50	50
185	46	46	47	47	48
190	43	44	45	45	46
195	42	42	43	43	43
200	40	41	41	41	42

Design procedure

1. Calculate the ultimate compressive force due to the axial loading

   $$F_c = \gamma_{f_i} W_i + \gamma_{f_d} W_d \quad \text{(kN)}$$

   $$F_c = 1{\cdot}6 W_i + 1{\cdot}4 W_d$$

   where $W_{i,d}$ = the characteristic imposed and dead axial loads and $\gamma_{f_{i,d}}$ = the load factors from Table 2, BS 5950.
2. Establish the effective length L_E for the column about both the XX and YY axes (see Figure 2.9 and Table 1.4):

   $$L_E = k \times L_{X,Y}$$

   where k = the effective length factor and L = the centre-to-centre distance between the points of effective restraint for the XX and YY axes.
3. Calculate the slenderness λ:

   $$\lambda_X = L_{EX}/r_X$$

   where r = the radius of gyration about the XX and YY axes,

   $$\lambda_Y = L_{EY}/r_Y$$

   The larger of the two values governs.

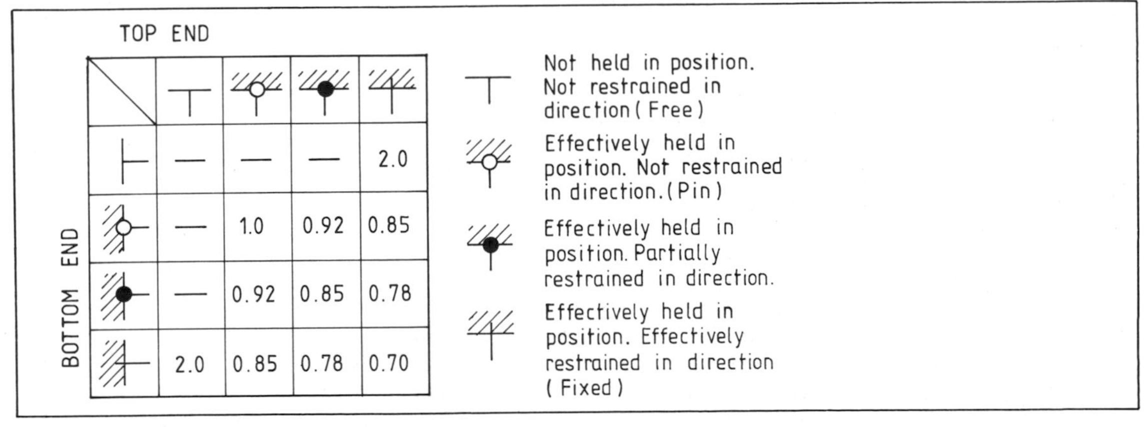

Figure 2.9 Effective length factor, k (nominal effective length $L_E = k \times L$). For stanchions in single storey buildings of simple construction refer to Appendix D of the code. For discontinuous angle struts see Figure 2.10 or refer to Table 28 of the code. When using the above k factor guide the method should be used for both the XX and YY axes in order to identify the critical value of λ

Check that $\lambda \leq 180$. If $\lambda > 180$ refer to clause 4.7.3.2 of the code.

4. Check the section classification. This need only be carried out for Universal Beam sections, as none of the Universal Columns or joists are slender under axial load only. Calculate d/t, where d = the depth of the web between fillets and t = the web thickness. The section is slender if:

$$\frac{d}{t} > 39$$

If the section is slender then the load-carrying capacity will be affected due to the influence of local buckling. In such cases the code requires a reduction in the design strength p_y calculated as follows:

$$p_{y_r} = \frac{39^2 \times 275}{(d/t)^2}$$

5. Obtain the compressive strength p_c. From Table 2.10 establish the appropriate column for the critical axis. (Table 2.8 is derived from the Perry strut formula, the full details of which are given in Appendix C of the code.) The compression strength p_c depends on the slenderness λ of the gross section and the design strength p_y as follows:

(a) $d/t \leq 39$, λ_{max} and p_y
(b) $d/t > 39$, λ_{max} and p_{y_r}

Table 2.9 Deflection limit δ_{lim} on beams due to unfactored loads

Member	*Deflection limit* δ_{lim}
Cantilever	Length/180
Beam carrying plaster or other brittle finish	Span/360
All other beams	Span/200

6. Calculate the compressive resistance P_c:

$$P_c = A_g \times P_c$$

where A_g = the gross sectional area (refer to section properties table) and P_c = the compressive strength based on λ and p_y or p_{y_r} according to the d/t ratio (see Tables 2.10 and 2.8).

7. Now check the compressive resistance P_c against the applied axial load F_c:

If $F_c \leq P_c$ the section is satisfactory, but if

$F_c > P_c$ choose a larger section and repeat steps 2–7.

Table 2.10 Compressive strength tables selection

Type of section	*Thickness*	*Axis of buckling* XX	YY
Rolled I-section		Table 2.8(a)	Table 2.8(b)
Rolled H-section	≤ 40 mm	Table 2.8(b)	Table 2.8(c)
Rolled angle Rolled channel Rolled T-section Two rolled sections back to back Compound rolled sections		Buckling about any axis (Table 2.8(c))	

I-section: a section with a central web and two equal flanges where $D > 1{\cdot}2B$.
H-section: a section with a central web and two equal flanges where $D < 1{\cdot}2B$.

2.4 Discontinuous single- and double-angle struts subject to axial load only

For this section the following assumptions are made:

1. Grade 43 steel is used.
2. The requirements of clause 4.7.13.1 for back-to-back struts are satisfied.
3. The strut is discontinuous.

Design procedure

Single-angle struts

1. Calculate the ultimate axial compression force, F_c, to be resisted by the strut.
2. Establish the length L of the strut.
3. Calculate the slenderness λ of the strut (Figure 2.10). An extract from Table 28 of the code gives the values of λ to be calculated according to the number of fasteners used at the end connections. The greater of the values for λ governs.
4. Check the classification of the section. As with columns under axial loading, if the strut section is slender a reduced design strength, p_{y_r} is used. The angle is slender if

$$\frac{b}{t} \text{ or } \frac{d}{t} > 15$$

where b = length of the shorter leg,
d = length of the longer leg,
t = thickness of the legs,

and $$\frac{b+d}{t} > 23$$

If the above values are less than 15 and 23, respectively, then the section is not slender and

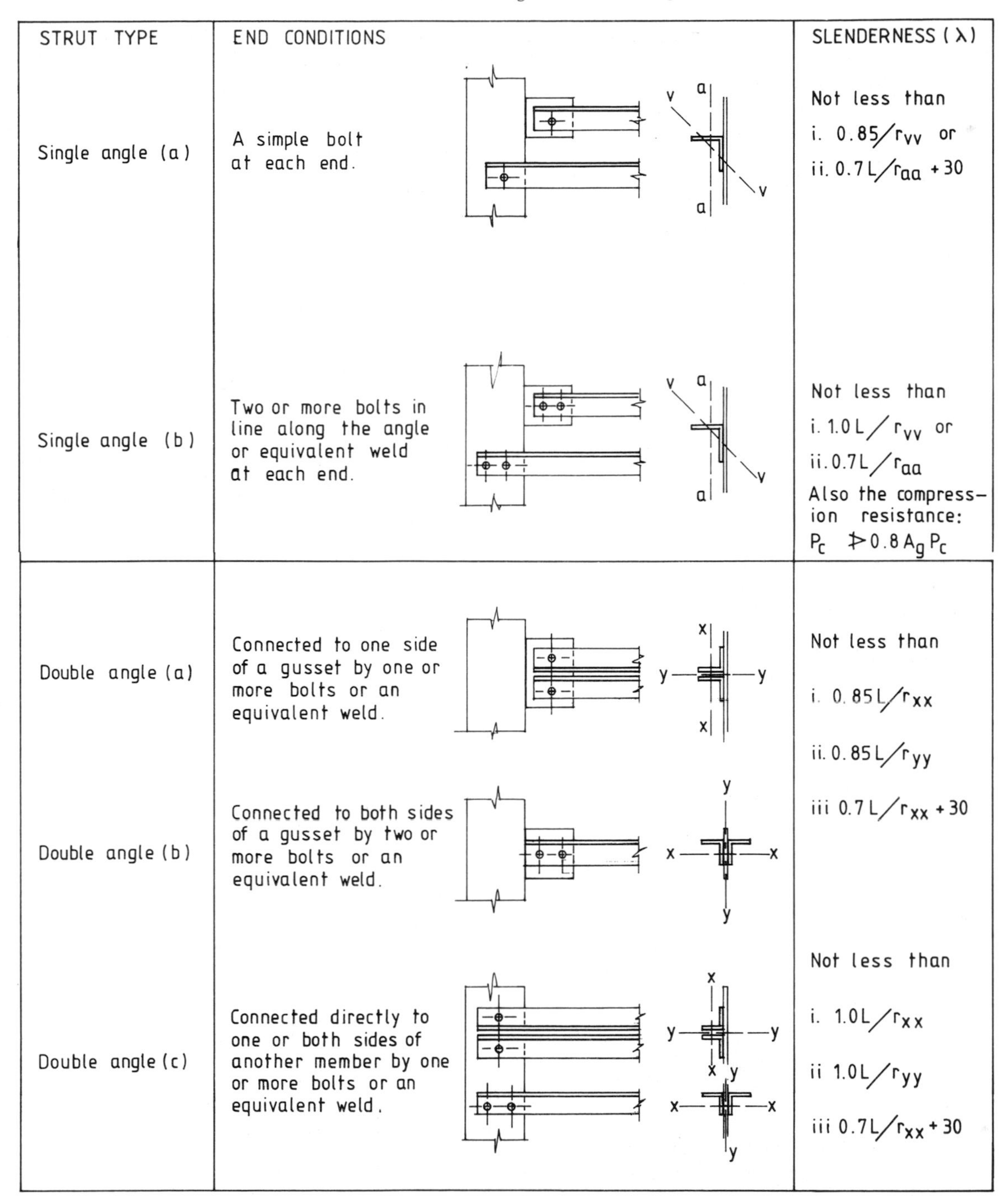

STRUT TYPE	END CONDITIONS	SLENDERNESS (λ)
Single angle (a)	A simple bolt at each end.	Not less than i. $0.85/r_{vv}$ or ii. $0.7L/r_{aa} + 30$
Single angle (b)	Two or more bolts in line along the angle or equivalent weld at each end.	Not less than i. $1.0L/r_{vv}$ or ii. $0.7L/r_{aa}$ Also the compression resistance: $P_c \ngtr 0.8 A_g p_c$
Double angle (a)	Connected to one side of a gusset by one or more bolts or an equivalent weld.	Not less than i. $0.85L/r_{xx}$ ii. $0.85L/r_{yy}$ iii $0.7L/r_{xx} + 30$
Double angle (b)	Connected to both sides of a gusset by two or more bolts or an equivalent weld.	
Double angle (c)	Connected directly to one or both sides of another member by one or more bolts or an equivalent weld.	Not less than i. $1.0L/r_{xx}$ ii $1.0L/r_{yy}$ iii $0.7L/r_{xx} + 30$

Figure 2.10 Angle end conditions

the design strength p_y from Table 2.1 is used. If the angle is slender calculate the reduction factor based on the least of

$$\text{(a)}\ \frac{11}{(d/t) - 4},\ \text{(b)}\ \frac{11}{(b/t) - 4},$$

$$\text{(c)}\ \frac{19}{(b + d)/t - 4}$$

and the reduced design strength

$$p_{y_r} = p_y \times \text{reduction factor}$$

5. Obtain the compressive strength p_c. From Table 2.8(c) using the appropriate value of λ, p_y or p_{y_r} read off p_c.
6. Calculate the compressive resistance P_c.

(a) End connection through one leg by two or more fasteners:

$P_L = A_g \times p_c$

where A_g = the gross sectional area (refer to section properties table).

(b) End connection through one leg by one fastener:

$P_c = 0{\cdot}8A_g \times p_c$

7. Now check the compressive resistance P_c against the applied ultimate axial load F_c.

If $F_c \leqslant P_c$ the section is satisfactory, but if $F_c > P_c$ choose a larger section and repeat steps 3–7.

Double-angle compound struts

1. Calculate the ultimate axial compression force F_c.
2. Establish the length L of the strut.
3. Calculate the slenderness λ which is the greater of

(a) $K_L \dfrac{L}{r_{XX}}$

(b) $\lambda_b = (\lambda_m^2 + \lambda_c^2)^{1/2}$

where $\lambda_m = K_L \dfrac{L}{r_{YY}}$ and $\lambda_c = \dfrac{L_c}{r_V} \not> 50$

(c) $0{\cdot}7\dfrac{L}{r_{XX}} + 30$

Note: K_L = 0·85 for struts connected to a gusset,
= 1·00 for struts connected to a member,

r_{XX} = the radius of gyration of the compound section about the XX axis,

r_{YY} = the radius of gyration of the compound section about the YY axis,

r_v = the minimum radius of gyration of a single angle,

L_c = $L/3$ (see clause 4.7.13.1(e) of the code).

4. Check the classification of the compound section. The section is slender if

b/t or $d/t > 15$

where b and d are the leg lengths of a single angle and t is the leg thickness. If b/t or $d/t \leqslant 15$ use p_y from Table 2.1, or if b/t or $d/t > 15$ reduce p_y from Table 2.1 by the lesser of the following factors for the reduced design strength, p_{y_r}:

(a) $\dfrac{11}{(d/t) - 4}$ or (b) $\dfrac{19}{(b + d)/t - 4}$

i.e. $p_{y_r} = p_y \times$ reduction factor.

5. Obtain the compressive strength, p_c. From Table 2.8(c) using the appropriate value of λ and p_y or p_{y_r} read off p_c.
6. Calculate the compressive resistance P_c.

$P_c = A_g \times p_c$

where A_g = the gross sectional area of the compound section.

7. Check the compressive resistance P_c against F_c. If $F_c \leqslant P_c$ the section is satisfactory, but if $F_c > P_c$ choose a larger section and repeat stages 3–7.

2.5 A cased column subject to axial load only

1. Check that the section of the cased column complies with the general requirements as shown in Figure 2.8.
2. Complete steps 1 and 2 as for the uncased section.
3. Calculate the slenderness λ for both axes, i.e.

(a) XX, $\lambda = \dfrac{L_E}{r_X}$

where r_X = the radius of gyration of the steel section about its XX axis,

(b) YY, $\lambda = \dfrac{L_E}{r_Y}$

where $r_Y = 0{\cdot}2b_c \not> 0{\cdot}2(B + 150)$ mm. Also if r_Y for the steel section is greater than r_Y for the cased one then r_Y for the steel section may be used.

4. Obtain the compressive strength p_c of the steel section. From Table 2.10 establish the appropriate column ((a), (b) or (c)) for the critical axis in Table 2.8.
5. Calculate the compressive resistance P_c and the short strut capacity P_{cs}:

$$P_c = \left(A_g + 0{\cdot}45\,\frac{f_{cu}}{p_y}\,A_c\right)p_c$$

$$P_{cs} = \left(A_g + 0{\cdot}25\,\frac{f_{cu}}{p_y}\,A_c\right)p_y$$

where A_g = the gross sectional area of the steel section,

A_c = the gross sectional area of the concrete, i.e. $A_c = d_c \times b_c - A_g$ and $d_c \not> D + 150$ mm, $b_c \not> B + 150$ mm,

f_{cu} = the 28-day characteristic strength of the concrete (but $f_{cu} \ngtr 40\ N/mm^2$),
p_c = the compressive strength of the steel section,
p_y = the design strength of the steel.

6. Check the compressive resistance P_c and the short strut capacity P_{cs} against F_c. The section is satisfactory if

$$F_c \leqslant P_c \text{ or if } P_c > P_{cs}$$

Then $F_c \leqslant P_{cs}$

2.6 Columns in simple multi-storey construction

For this section the following assumptions are made:

1. The moments applied to the column are due only to the eccentricity of the connections.
2. The design assumptions made for 'Columns subject to axial load only' (see page 31) apply to this section.

Design procedure

Clause 4.8.3.3.1 states that the following relationship is satisfied:

$$\frac{F_c}{A_g p_c} + \frac{mM_X}{M_b} + \frac{mM_Y}{p_y Z_Y} \leqslant 1{\cdot}0$$

1. $\dfrac{F_c}{A_g p_c}$

where F_c = the applied ultimate axial load,
A_g = the gross section area of the section,
p_c = the compressive strength.

To calculate F_c and p_c follow the design procedure for 'Columns subject to axial load only'.

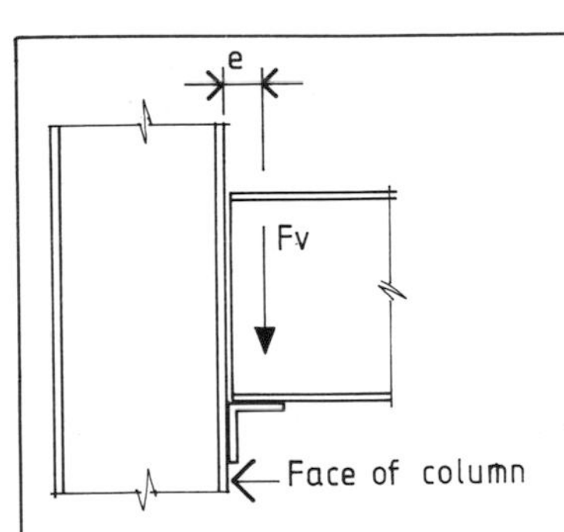

1) *Beam to column connection via column web or flange*

e' = eccentricity of load from the face of the steel column

e' = the greater of 100 mm or $b_1/2$

where b_1 is the stiff bearing *see* Figure 2.6.

Moment due to eccentricity, e

$M_e = F_v \times e$

where for 1) beam to column flange connection $e = e' + D/2$

2) beam to column web connection $e = e' + t/2$

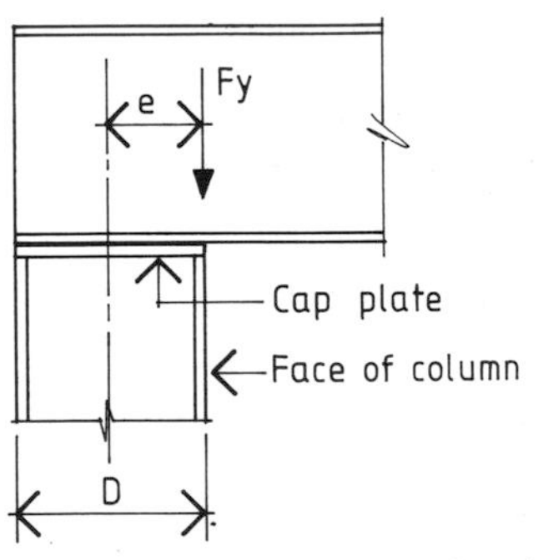

$M_e = F_v \times e$

where $e = D/2$

2) *Beam to column connection via a cap plate*

3) *Roof truss supported on a cap plate*

The eccentricity, e, may be neglected provided simple connections are used which do not develop significant moments.

hence $M_e = 0$

Figure 2.11 Moment and eccentricity of beam-to-column connection

2. $\dfrac{mM_X}{m_b}$

where m = the equivalent uniform moment factor (= 1·0),
M_X = the applied moment about the XX axis due to eccentric connections,
M_b = the buckling resistance moment capacity about the XX axis.

(a) Calculate the moment M_e due to the eccentricity of the connections about the XX axis. The moment M_X = the algebraic sum of the moments M_e (see Figure 2.11 for values of eccentricity e and M_e.) The moment M_X is now distributed between the upper and lower columns at the connection according to the procedure shown in Figure 2.12.
(b) Calculate the buckling resistance moment M_b:

$$M_b = S_X p_b$$

where S_X = plastic modulus about XX and p_b = the bending strength. For p_b, calculate the equivalent slenderness:

$$\lambda_{LT} = 0{\cdot}5\left(\frac{L}{r_Y}\right)$$

where $L = L_2$ in Figure 2.12 and r_Y = radius of gyration about the YY axis and from Table 2.1 the design strength p_y. p_b is now read off from Table 2.5 for the above values of λ_{LT} and p_y.

3. $\dfrac{mM_Y}{p_y Z_Y}$

where m = the equivalent moment factor (= 1·0), M_Y = the applied moment about the YY axis due to eccentric connections (M_Y is calculated as shown for M_X in 2(a) above), p_y = the design strength of the steel and Z_Y = the elastic modulus about YY.

4. Check the ratios of applied axial load and moments and section capacity:

$$\frac{F_c}{A_g p_c} + \frac{mM_X}{M_b} + \frac{mM_Y}{p_y Z_Y} \leqslant 1{\cdot}0$$

If the above is greater than 1·0 choose a larger section and repeat steps 1–4.

2.7 Simple tension members

For this section the following assumptions are made:

1. The member is formed from one of the following sections:

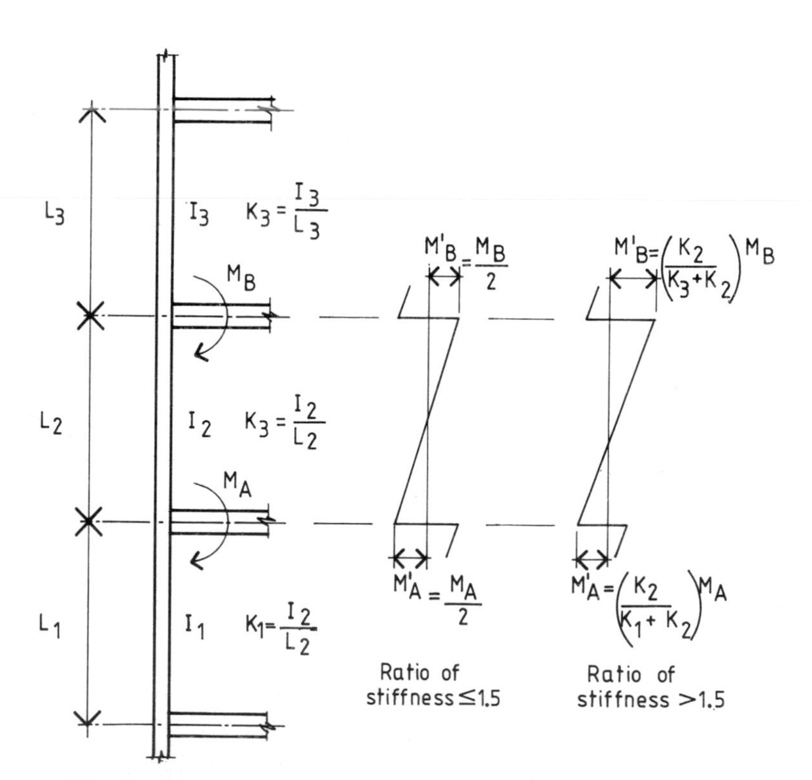

Figure 2.12 Distribution of column net moments by relative stiffness

(a) A flat bar,
(b) A single angle, or
(c) Back-to-back double angles.
2. Grade 43 steel is used.

Design procedure

Clause 4.6.1 of the code states that the tensile capacity P_t of a member should be taken from

$$P_t = A_e \times p_y$$

where A_e = the effective area.

1. For flat bars:

(a) Calculate the applied tensile force F_T.
(b) Calculate the net area A_n based on the gross area A_g, minus the allowance. (For bolt holes, see Figure 2.13.)
(c) Calculate the effective area A_e:

$$A_e = K_e \times A_n$$

where $K_e = 1{\cdot}2$ for Grade 43 steel but $A_e \leqslant A_g$.
(d) Calculate the tensile capacity $P_t = A_e \times p_y$.
(e) Check $F_t \leqslant P_t$.

2. For single angles connected through one leg only:
(a) Calculate the applied tensile force F_T.
(b) Calculate the net area of the connected leg a_1.
(c) Calculate the gross area of the unconnected leg a_2.
(d) Calculate the effective area A_e of the angle

$$A_e = a_1 + a_2\left(\frac{3a_1}{3a_1 + a_2}\right)$$

(e) Follow steps (d) and (e) of 1 above.
3. Back-to-back double angles connected to one side of a gusset or member should be designed as for 2 above or, if compound (see clause 4.6.3.2 of the code), as follows.
(a) Calculate the applied tensile force F_t.
(b) Calculate the net area of the two connected legs a_1.
(c) Calculate the gross area of the two unconnected legs a_2.
(d) Calculate the effective area A_e of the two angles

$$A_e = a_1 + a_2\left(\frac{Sa_1}{Sa_1 + a_2}\right)$$

(e) Follow steps (d) and (e) of 1 above.

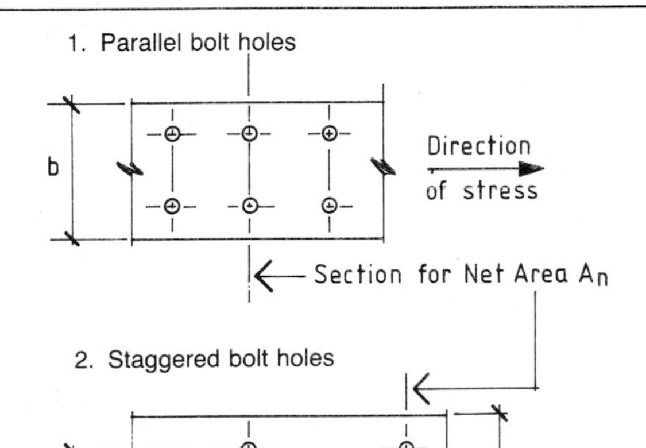

The net area A_n is taken as

$$A_n = A_g - \sum td$$

where A_g = the gross area, t is the member thickness, d is the diameter of the bolt holes.

The net area A_n is taken as

$$A_n = A_g - \sum td + \frac{S_p^2 t}{4g}$$

where A_g, t and d are as above.

3. Gross area A_g

(a) Flat bars $A_g = b \times t$
(b) Leg of angle $A_g = t(d - t/2)$
where b = width of the bar;
d = the leg length of the angle;
t = the thickness of the bar or leg of the angle

Figure 2.13 Gross and net areas of connected member

2.8 Simple connections

For this section the following assumptions are made:

1. The connection is designed to transfer direct loads only.
2. The bolts are to be one of the following:
   (a) Black hexagonal to BS 4190, Grade 4.6;
   (b) Precision hexagonal to BS 3692, Grade 8.8; or
   (c) High-strength friction-grip (HSFG) to BS 4395, General Grade.
3. Welds are formed using Grade E43 electrodes to BS 639.
4. Connected members, plates and cleats are formed from Grade 43 steel.
5. All fasteners are in clearance holes (i.e. nominal bolt diameter plus 2 mm).

Governing capacity of Grades 4.6 and 8.8 ordinary bolts in shear

The capacity of an ordinary bolt is the least value of:

1. Shear capacity in single or double shear (see Figure 2.14);
2. Bearing capacity of the bolt;
3. Bearing capacity of the connected ply.

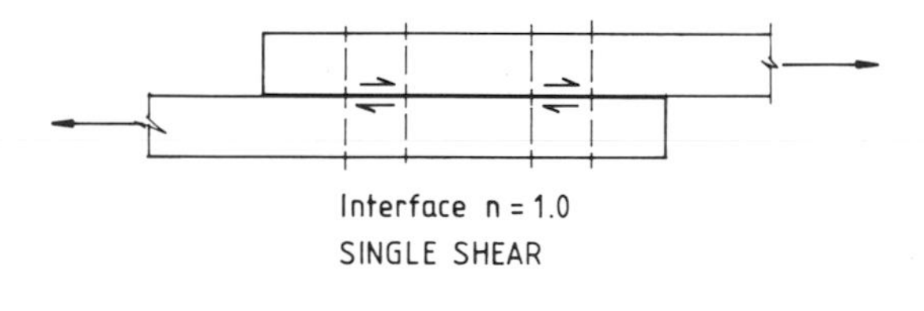

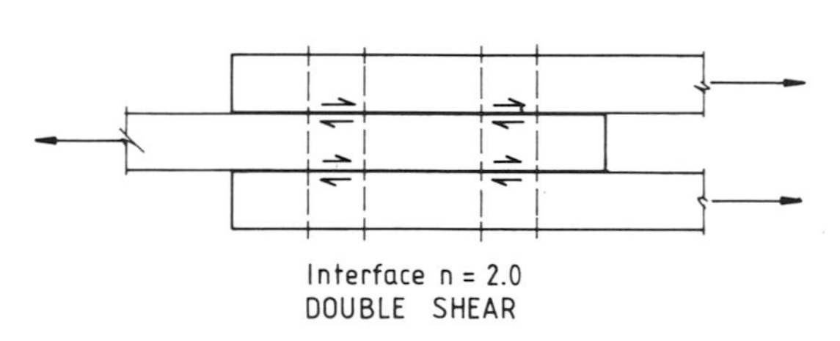

Figure 2.14

Shear capacity

$$P_s = p_s A_s n$$

where p_s and A_s are obtained from Table 2.11 and n = number of interfaces.

Bearing capacity of the bolt

$$P_{bb} = dtp_{bb}$$

where d = the nominal diameter of the bolt, t = the thickness of the connected ply and p_{bb} for Grade 4.6 bolts = 460 N/mm^2 and Grade 8.8 bolts = 970 N/mm^2.

Bearing capacity of connected ply

$$P_{bs} = dtp_{bs} \leqslant \tfrac{1}{2}etp_{bs}$$

where d and t are as above, p_{bs} for Grade 43 steel = 460 N/mm^2 and e = the end distance as defined in Figure 2.15.

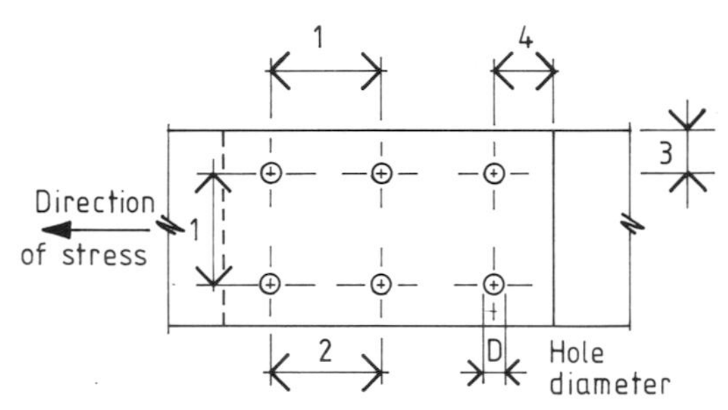

Figure 2.15 Rules for spacing of bolts and edge distances. (1) Minimum spacing between centres of bolts $\nless 2.5 \times$ the nominal diameter of the bolt; (2) maximum spacing between centres of bolts not exposed to corrosive influences $\ngtr 14t$, where t is the thickness of the thinner ply; (3) minimum edge distance (a) $\nless 1.25D$ for rolled edges, (b) $\nless 1{\cdot}40D$ for sheared edges (4) minimum end distance $\nless 1{\cdot}4D$

Governing capacity of general grade HSFG bolts resisting shear

The capacity of an HSFG bolt is the lesser value of:

1. Slip resistance, or
2. Bearing capacity of the connected ply.

Slip resistance in single or double shear (Figure 2.14)

$$P_{SL} = 1{\cdot}1 k_s \mu P_0 n$$

where P_0 = the proof load of the bolt from Table 2.11, $k_s = 1{\cdot}0$, μ = the slip factor (= 0·45) and n = number of interfaces.

Bearing capacity of connected ply

$$P_{bg} = dtp_{bg} \leqslant \tfrac{1}{3}etp_{bg}$$

where d = nominal diameter of the bolt, t = the thickness of the connected ply, e = the end distance as defined in Figure 2.15 and p_{bg} for Grade 43 steel = 825 N/mm^2.

Table 2.11 Bolt and fillet weld capacities

4·6 bolts

Diameter of bolt (mm)	*Tensile stress area* A_s *(mm²)*	*Tensile cap* P_t *(kN)*	*Shear value* P_s *Single shear (kN)*	*Double shear (kN)*
12	84·3	16·4	13·5	27·0
16	157	30·6	25·1	50·2
20	245	47·8	39·2	78·4
22	303	59·1	48·5	97·0
24	353	68·8	56·5	113
27	459	89·5	73·4	147
30	561	109	89·8	180

8·8 bolts

Diameter of bolt (mm)	*Tensile stress area* A_s *(mm²)*	*Tensile cap* P_t *(kN)*	*Shear value* P_s *Single shear (kN)*	*Double shear (kN)*
12	84·3	37·9	31·6	63·2
16	157	70·7	58·9	118
20	245	110	91·9	184
22	303	136	114	227
24	353	159	132	265
27	459	207	172	344
30	561	252	210	421

General Grade HSFG bolts

Diameter of bolt (mm)	*Proof load of bolt* P_0 *(kN)*	*Tensile cap* P_t *(kN)*	*Slip value* P_{SL} *Single shear (kN)*	*Double shear (kN)*
12	49·4	44·5	24·5	48·9
16	92·1	82·9	45·6	91·2
20	144	130	71·3	143
22	177	159	87·6	175
24	207	186	102	205
27	234	211	116	232
30	286	257	142	283

Fillet weld capacities with Grade E43 electrodes

Leg length (mm)	*Throat thickness (mm)*	*Capacity at 215 N/mm² (kN/mm)*	*Leg length (mm)*	*Throat thickness (mm)*	*Capacity at 215 N/mm² (kN/mm)*
3·0	2·1	0·452	112·0	8·4	1·81
4·0	2·8	0·602	15·0	10·5	2·26
5·0	3·5	0·753	18·0	12·6	2·71
6·0	4·2	0·903	20·0	14·0	3·01
8·0	5·6	1·2	22·0	15·4	3·31
10·0	7·0	1·51	25·0	17·5	3·76

Capacity of Grades 4.6 and 8.8 ordinary bolts in combined tension and shear

The following relationship must be satisfied:

$$\frac{F_s}{P_s} + \frac{F_t}{P_t} \leq 1{\cdot}4$$

where F_s is the applied shear, P_s is the shear capacity of the bolt, F_t is the applied tension and P_t is the tension capacity (see Table 2.11 for values).

Capacity of HSFG bolts in combined tension and shear

The following relationship must be satisfied:

$$\frac{F_s}{P_{SL}} + 0{\cdot}8\,\frac{F_t}{P_t} \leq 1{\cdot}0$$

where F_s is the applied shear, P_{SL} is the slip resistance, F_t is the applied tension and $P_t = 0{\cdot}9\ P_0$, where $P_0 =$ the proof load.

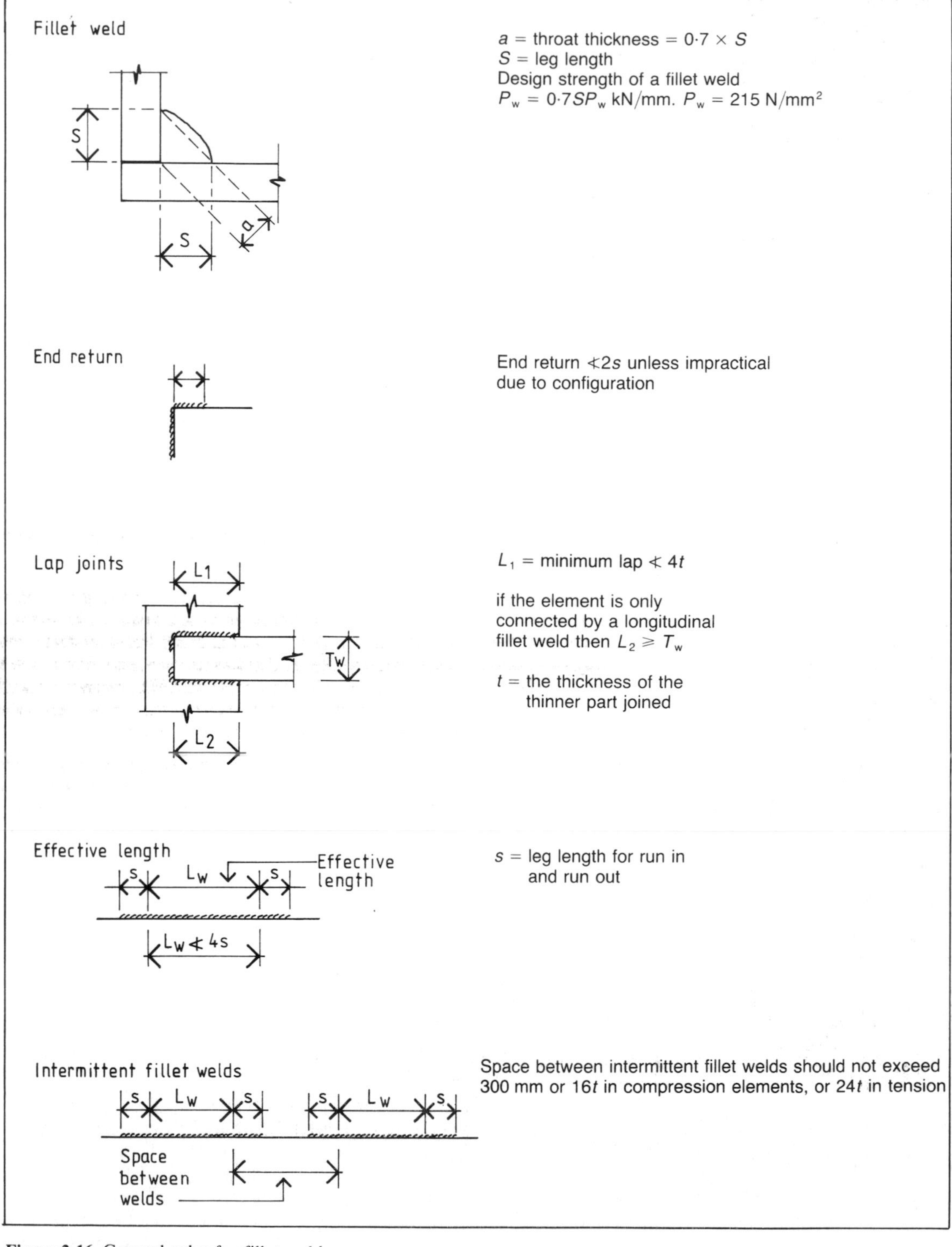

Figure 2.16 General rules for fillet welds

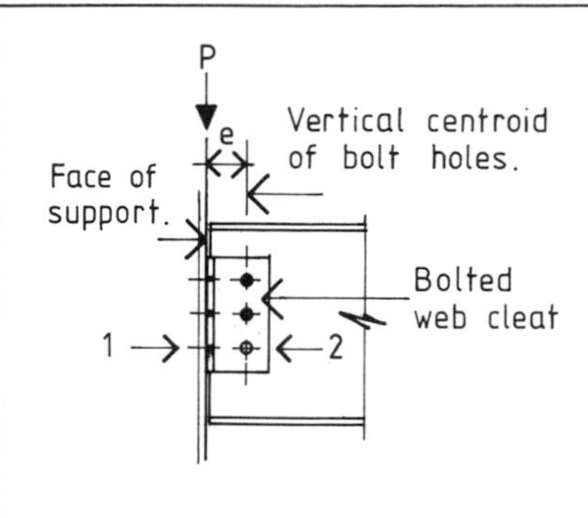

Bolts (1) carry vertical shear only (page 38). Bolts (2) carry vertical shear and shear due to eccentricity of P. (Table 2.11 and Figure 2.18)

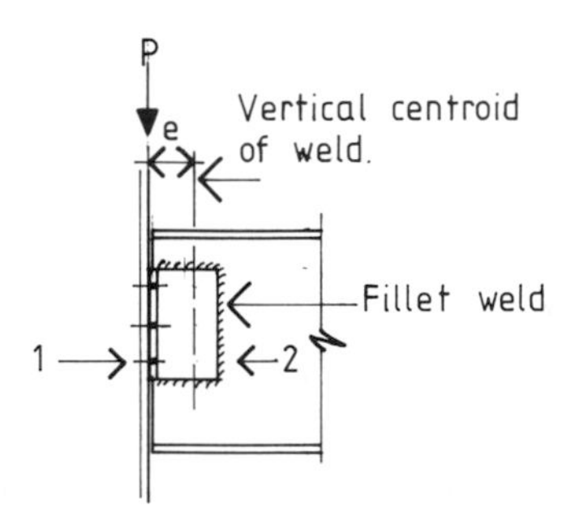

Bolts (1) carry vertical shear only (page 38). Weld (2) carries vertical shear and shear due to eccentricity of P. (Table 2.11 and Figures 2.16 and 2.18)

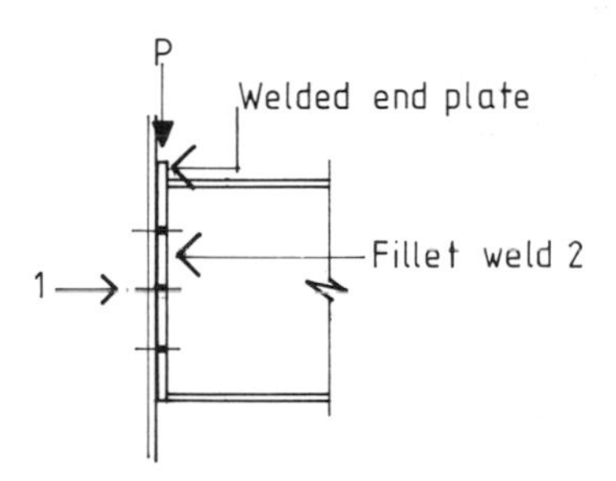

Bolts (1) carry vertical shear only (page 38). Weld (2) carries vertical shear only. (Tables 2.11 and 2.12)

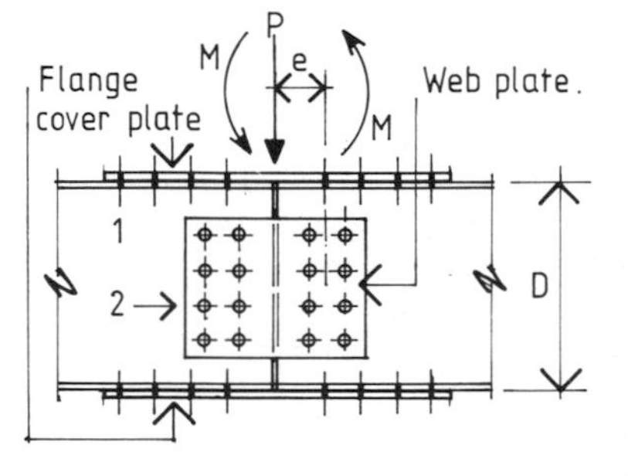

Bolts (1) carry a shear force

$F_T = M/D$ (page 38).

Bolts (2) carry vertical shear and shear due to eccentricity of P. (page 38 and Figure 2.18)
Normally HSFG bolts are used for this type of connection
Flange cover plate to be checked for tension due to F_T. Web cover plate to be checked for bending and shear

Figure 2.17 Typical beam connections

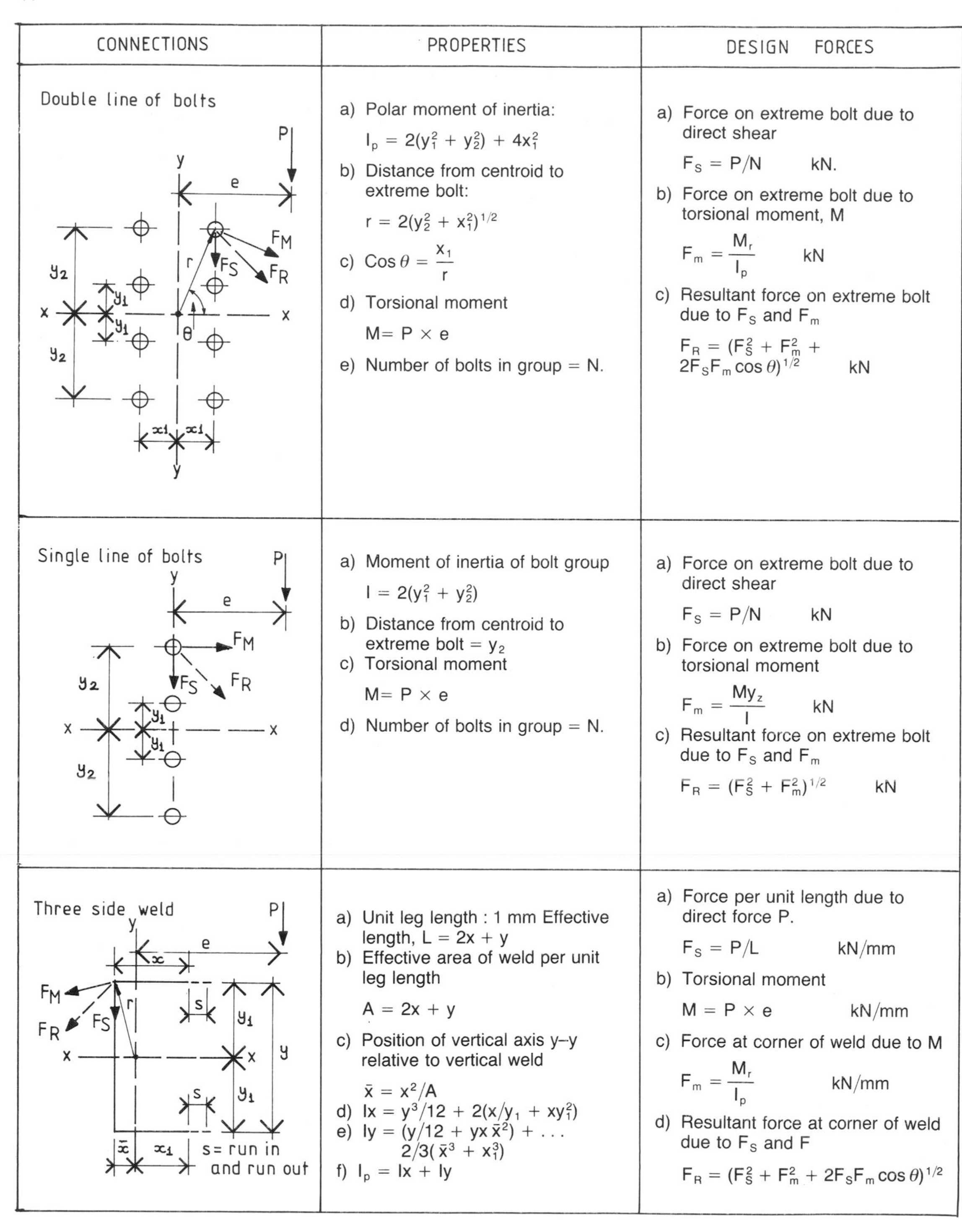

CONNECTIONS	PROPERTIES	DESIGN FORCES
Double line of bolts	a) Polar moment of inertia: $I_p = 2(y_1^2 + y_2^2) + 4x_1^2$ b) Distance from centroid to extreme bolt: $r = 2(y_2^2 + x_1^2)^{1/2}$ c) $\cos\theta = \frac{x_1}{r}$ d) Torsional moment $M = P \times e$ e) Number of bolts in group = N.	a) Force on extreme bolt due to direct shear $F_S = P/N$ kN. b) Force on extreme bolt due to torsional moment, M $F_m = \frac{M_r}{I_p}$ kN c) Resultant force on extreme bolt due to F_S and F_m $F_R = (F_S^2 + F_m^2 + 2F_SF_m\cos\theta)^{1/2}$ kN
Single line of bolts	a) Moment of inertia of bolt group $I = 2(y_1^2 + y_2^2)$ b) Distance from centroid to extreme bolt = y_2 c) Torsional moment $M = P \times e$ d) Number of bolts in group = N.	a) Force on extreme bolt due to direct shear $F_S = P/N$ kN b) Force on extreme bolt due to torsional moment $F_m = \frac{My_z}{I}$ kN c) Resultant force on extreme bolt due to F_S and F_m $F_R = (F_S^2 + F_m^2)^{1/2}$ kN
Three side weld s = run in and run out	a) Unit leg length : 1 mm Effective length, $L = 2x + y$ b) Effective area of weld per unit leg length $A = 2x + y$ c) Position of vertical axis y–y relative to vertical weld $\bar{x} = x^2/A$ d) $I_x = y^3/12 + 2(x/y_1 + xy_1^2)$ e) $I_y = (y/12 + yx\bar{x}^2) + \ldots$ $2/3(\bar{x}^3 + x_1^3)$ f) $I_p = I_x + I_y$	a) Force per unit length due to direct force P. $F_S = P/L$ kN/mm b) Torsional moment $M = P \times e$ kN/mm c) Force at corner of weld due to M $F_m = \frac{M_r}{I_p}$ kN/mm d) Resultant force at corner of weld due to F_S and F $F_R = (F_S^2 + F_m^2 + 2F_SF_m\cos\theta)^{1/2}$

Figure 2.18 Eccentrically loaded connections

Table 2.12 Recommended back marks
H- and I-sections

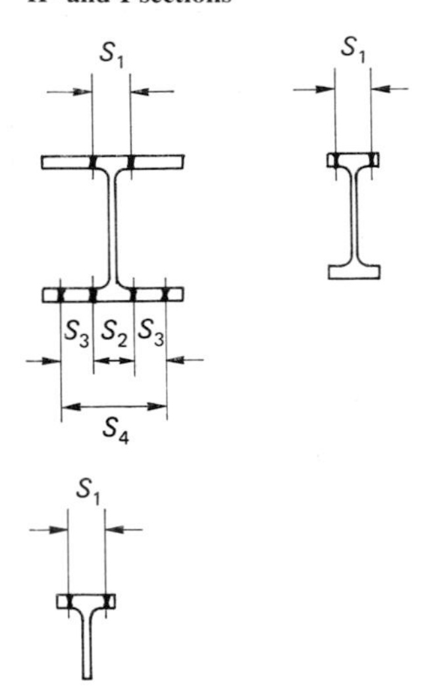

Nominal flange width (mm)	*Spacing of holes* S_1 *(mm)*	S_2 *(mm)*	S_3 *(mm)*	S_4 *(mm)*	*Maximum dia. of rivet or bolt (mm)*	*b (mm)*
419 to 368	140	140	75	290	24	362
330 and 305	140	120	60	240	24	312
330 and 305	140	120	60	240	20	300
292 to 203	140				24	212
190 to 165	90				24	162
152	90				20	150
146 to 127	70				20	130
102	54				12	98
89	50					
76	40					
64	34					
51	30					

Note that the actual flange width for a universal section may be less than the nominal size and that the difference may be significant in determining the maximum diameter.

The dimensions S_1 and S_2 have been selected for normal conditions but adjustments may be necessary for relatively large-diameter fastenings or particularly heavy masses of serial size.

Standard Angles

These angles are those metric sizes selected, from the full list recommended by the ISO, as British Standard Metric Angles. They replaced the Imperial sizes completely from 1 January 1973.

Note that HSFG bolts may require adjustments to the back marks shown due to the larger nut and washer dimensions.

Inner gauge lines are for normal conditions and may require adjustment for large diameters of fasteners or thick members.

Outer gauge lines may require consideration in relation to a specified edge distance.

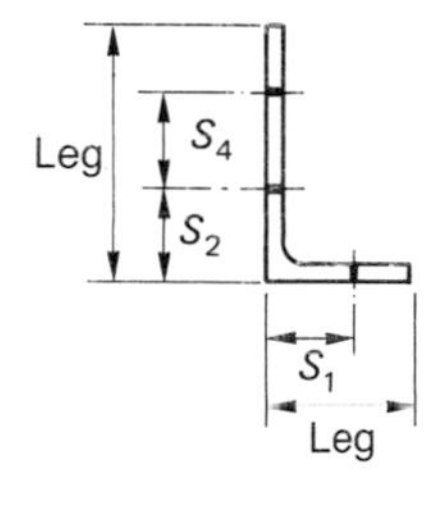

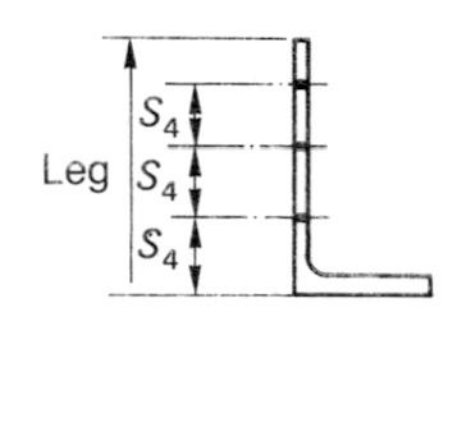

Nomial leg length (mm)	*Spacing of holes* S_1 *(mm)*	S_2 *(mm)*	S_3 *(mm)*	S_4 *(mm)*	S_5 *(mm)*	S_6 *(mm)*	*Maximum diameter of bolt or rivet* S_1 *(mm)*	S_2 *and* S_3 *(mm)*	$S_4 S_5$ *and* S_6 *(mm)*
200		75	75	55	55	55		30	20
150		55	55					20	
125		45	50					20	
120		45	50					16	
100	55						24		
90	50						24		
80	45						20		
75	45						20		
70	40						20		
65	35						20		
60	35						16		
50	28						12		
45	25								
40	23								
30	20								
25	15								

Safe load tables

The actual values for the capacity in shear of 4·6, 8·8 and HSFG bolts may be directly read off the safe load tables shown in Table 2.11, as may their tensile capacity P_t.

Design

It is not possible to give a useful step-by-step guide for the design of simple connections as the required checks depend on the configuration of the connection. However, certain general rules can be given to aid the initial layout of the connection:

1. Bolts used for structural connections should not be less than 16 mm diameter.
2. Fillet welds used for structural connections should not be less than 6 mm.
3. All bolts in a connection should be one size and grade.
4. All welds in a connection should be one size and grade.
5. Both bolt and weld configurations should be symmetrical.
6. The initial sizing of plates and angles through which bolts pass may be achieved by using the rules given in Figure 2.15 and Table 2.12.

Figure 2.17 shows typical simple connections identifying the checks to be carried out on the various elements.

2.9 A column base plate subject to axial compressive load only (Figure 2.19)

1. Establish the factored axial compressive load, F_c.
2. Establish the design strength of the plate p_{y_p} (p_{y_p} is obtained from Table 2.1). It is unlikely that the thickness of the plate will be less than 16 mm; therefore assume an initial value for $p_{y_p} = 265\ \text{N/mm}^2$.
3. It is assumed that the base plate is supported on a concrete base and the bearing strength of the concrete is taken as 40% of the characteristic 28-day cube strength (i.e. $0{\cdot}4f_{cu}$).
4. Calculate the area of base plate required:

$$A_{\text{Preqd}} = \frac{F_c}{0{\cdot}4f_{cu}}$$

5. Assuming the sides of the base plate are equal, then:

$$B_p = D_p = (A_{\text{Preqd}})^{1/2}$$

As the calculated value of B_p and D_p is rarely an exact number, they should now be rounded up to practical dimensions. If, for example, $A_{\text{Preqd}} = 684\,000\ \text{mm}^2$ then $B_p = D_p = 827$ mm. Therefore use 850 × 850 mm plate: $A_p = 850 \times 850 = 722\,500\ \text{mm}^2$.

6. Calculate the pressure on the underside of the plate:

$$w = \frac{F_c}{A_p}$$

7. Calculate the minimum thickness of the base plate:

$$t_{\min} = \left[\frac{2{\cdot}5}{p_{y_p}} w(a^2 - 0{\cdot}3b^2)\right]^{1/2}$$

If $t_{\min} > 40$ mm reduce $p_{y_p} = 245\ \text{N/mm}^2$ and recalculate $t_{\min}$. The actual t should not be less in practice than the thickness of the flange of the supported member and must be a multiple of 5.

2.10 Section properties

Tables 2.13–2.19 give the properties of some steel sections. Full details and safe load tables may be obtained from BSC General Steels Group.

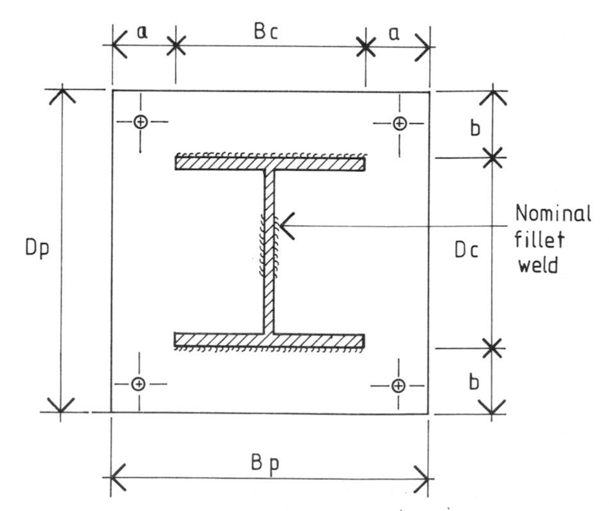

Figure 2.19 a = the greater projection and b = the lesser projection of the plate beyond the column

Table 2.13 Section properties of Universal Beams to BS4: Part 1

Designation		Depth of section	Width of section	Thickness		Root radius	Depth between fillets	Ratios for local buckling		Second moment of area		Radius of gyration		Elastic modulus		Plastic modulus		Buckling parameter	Torsional index	Warping constant	Torsional constant	Area of section
Serial size (mm)	Mass per metre (kg)	D (mm)	B (mm)	Web t (mm)	Flange T (mm)	r (mm)	d (mm)	Flange b/T	Web d/t	XX axis (cm⁴)	YY axis (cm⁴)	XX axis (cm)	YY axis (cm)	XX axis (cm³)	YY axis (cm³)	XX axis (cm³)	YY axis (cm³)	u	x	H (dm⁶)	J (cm⁴)	A (cm²)
914×419	388	920·5	420·5	21·5	36·6	24·1	799·1	5·74	37·2	719 000	45 400	38·1	9·58	15 600	2160	17 700	3340	0·884	26·7	88·7	1730	494
	343	911·4	418·5	19·4	32·0	24·1	799·1	6.54	41·2	625 000	39 200	37·8	9·46	13 700	1870	15 500	2890	0·883	30·1	75·7	1190	437
914×305	289	926·6	307·8	19·6	32·0	19·1	824·5	4·81	42·1	505 000	15 600	37·0	6·51	10 900	1010	12 600	1600	0·867	31·9	31·2	929	369
	253	918·5	305·5	17·3	27·9	19·1	824·5	5·47	47·7	437 000	13 300	36·8	6·42	9 510	872	10 900	1370	0·866	36·2	26·4	627	323
	224	910·3	304·1	15·9	23·9	19·1	824·5	6·36	51·9	376 000	11 200	36·3	6·27	8 260	738	9 520	1160	0·861	41·3	22·0	421	285
	201	903·0	303·4	15·2	20·2	19·1	824·5	7·51	54·2	326 000	9 430	35·6	6·06	7 210	621	8 360	983	0·853	46·8	18·4	293	256
838×292	226	850·9	293·8	16·1	26·8	17·8	761·7	5·48	47·3	340 000	11 400	34·3	6·27	7 990	773	9 160	1210	0·87	35·0	19·3	514	289
	194	840·7	292·4	14·7	21·7	17·8	761·7	6·74	51·8	279 000	9 070	33·6	6·06	6 650	620	7 650	974	0·862	41·6	15·2	307	247
	176	834·9	291·6	14·0	18·8	17·8	761·7	7·76	54·4	246 000	7 790	33·1	5·90	5 890	534	6 810	842	0·856	46·5	13·0	222	224
762×267	197	769·6	268·0	15·6	25·4	16·5	685·8	5·28	44·0	240 000	8 170	30·9	5·71	6 230	610	7 170	959	0·869	33·2	11·3	405	251
	173	762·0	266·7	14·3	21·6	16·5	685·8	6·17	48·0	205 000	6 850	30·5	5·57	5 390	513	6 200	807	0·864	38·1	9·38	267	220
	147	753·9	265·3	12·9	17·5	16·5	685·8	7·58	53·2	169 000	5 470	30·0	5·39	4 480	412	5 170	649	0·857	45·1	7·41	161	188
686×254	170	692·9	255·8	14·5	23·7	15·2	615·1	5·40	42·4	170 000	6 620	28·0	5·53	4 910	518	5 620	810	0·872	31·8	7·41	307	217
	152	687·6	254·6	13·2	21·0	15·2	615·1	6·06	46·6	150 000	5 780	27·8	5·46	4 370	454	5 000	710	0·871	35·5	6·42	219	194
	140	683·5	253·7	12·4	19·0	15·2	615·1	6·68	49·6	136 000	5 180	27·6	5·38	3 990	408	4 560	638	0·868	38·7	5·72	169	179
	125	677·9	253·0	11·7	16·2	15·2	615·1	7·81	52·6	118 000	4 380	27·2	5·24	3 480	346	4 000	542	0·862	43·9	4·79	116	160
610×305	238	633·0	311·5	18·6	31·4	16·5	537·2	4·96	28·9	208 000	15 800	26·1	7·22	6 560	1020	7 460	1570	0·886	21·1	14·3	788	304
	179	617·5	307·0	14·1	23·6	16·5	537·2	6·50	38·1	152 000	11 400	25·8	7·08	4 910	743	5 520	1140	0·886	27·5	10·1	341	228
	149	609·6	304·8	11·9	19·7	16·5	537·2	7·74	45·1	125 000	9 300	25·6	6·99	4 090	610	4 570	937	0·886	32·5	8·09	200	190
610×229	140	617·0	230·1	13·1	22·1	12·7	547·3	5·21	41·8	112 000	4 510	25·0	5·03	3 630	392	4 150	612	0·875	30·5	3·99	217	178
	125	611·9	229·0	11·9	19·6	12·7	547·3	5·84	46·0	98 600	3 930	24·9	4·96	3 220	344	3 680	536	0·873	34·0	3·45	155	160
	113	607·3	228·2	11·2	17·3	12·7	547·3	6·60	48·9	87 400	3 440	24·6	4·88	2 880	301	3 290	470	0·87	37·9	2·99	112	144
	101	602·2	227·6	10·6	14·8	12·7	547·3	7·69	51·6	75 700	2 910	24·2	4·75	2 510	256	2 880	400	0·863	43·0	2·51	77·2	129

Table 2.13 (*cont.*)

Designation: Serial size (mm)	Designation: Mass per metre (kg)	Depth of section D (mm)	Width of section B (mm)	Thickness: Web t (mm)	Thickness: Flange T (mm)	Root radius r (mm)	Depth between fillets d (mm)	Ratios for local buckling: Flange b/T	Ratios for local buckling: Web d/t	Second moment of area: XX axis (cm^4)	Second moment of area: YY axis (cm^4)	Radius of gyration: XX axis (cm)	Radius of gyration: YY axis (cm)	Elastic modulus: XX axis (cm^3)	Elastic modulus: YY axis (cm^3)	Plastic modulus: XX axis (cm^3)	Plastic modulus: YY axis (cm^3)	Buckling parameter u	Torsional index x	Warping constant H (dm^6)	Torsional constant J (cm^4)	Area of section A (cm^2)
533×210	122	544·6	211·9	12·8	21·3	12·7	476·5	4·97	37·2	76 200	3 390	22·1	4·67	2 800	320	3 200	501	0·876	27·6	2·32	180	156
	109	539·5	210·7	11·6	18·8	12·7	476·5	5·60	41·1	66 700	2 940	21·9	4·60	2 470	279	2 820	435	0·875	30·9	1·99	126	139
	101	536·7	210·1	10·9	17·4	12·7	476·5	6·04	43·7	61 700	2 690	21·8	4·56	2 300	257	2 620	400	0·874	33·1	1·82	102	129
	92	533·1	209·3	10·2	15·6	12·7	476·5	6·71	46·7	55 400	2 390	21·7	4·51	2 080	229	2 370	356	0·872	36·4	1·60	76·2	118
	82	528·3	208·7	9·6	13·2	12·7	476·5	7·91	49·6	47 500	2 010	21·3	4·38	1 800	192	2 060	300	0·865	41·6	1·33	51·3	104
457×191	98	467·4	192·8	11·4	19·6	10·2	407·9	4·92	35·8	45 700	2 340	19·1	4·33	1 960	243	2 230	378	0·88	25·8	1·17	121	125
	89	463·6	192·0	10·6	17·7	10·2	407·9	5·42	38·5	41 000	2 090	19·0	4·28	1 770	217	2 010	338	0·879	28·3	1·04	90·5	114
	82	460·2	191·3	9·9	16·0	10·2	407·9	5·98	41·2	37 100	1 870	18·8	4·23	1 610	196	1 830	304	0·877	30·9	0·923	69·2	105
	74	457·2	190·5	9·1	14·5	10·2	407·9	6·57	44·8	33 400	1 670	18·7	4·19	1 460	175	1 660	272	0·876	33·9	0·819	52·0	95·0
	67	453·6	189·9	8·5	12·7	10·2	407·9	7·48	48·0	29 400	1 450	18·5	4·12	1 300	153	1 470	237	0·873	37·9	0·706	37·1	85·4
457×152	82	465·1	153·5	10·7	18·9	10·2	407·0	4·06	38·0	36 200	1 140	18·6	3·31	1 560	149	1 800	235	0·872	27·3	0·569	89·3	104
	74	461·3	152·7	9·9	17·0	10·2	407·0	4·49	41·1	32 400	1 010	18·5	3·26	1 410	133	1 620	209	0·87	30·0	0·499	66·6	95·0
	67	457·2	151·9	9·1	15·0	10·2	407·0	5·06	44·7	28 600	878	18·3	3·21	1 250	116	1 440	182	0·867	33·6	0·429	47·5	85·4
	60	454·7	152·9	8·0	13·3	10·2	407·7	5·75	51·0	25 500	794	18·3	3·23	1 120	104	1 280	163	0·869	37·5	0·387	33·6	75·9
	52	449·8	152·4	7·6	10·9	10·2	407·7	6·99	53·6	21 300	645	17·9	3·11	949	84·6	1 090	133	0·859	43·9	0·311	21·3	66·5
406×178	74	412·8	179·7	9·7	16·0	10·2	360·5	5·62	37·2	27 300	1 540	17·0	4·03	1 320	172	1 500	267	0·881	27·6	0·608	63·0	95·0
	67	409·4	178·8	8·8	14·3	10·2	360·5	6·25	41·0	24 300	1 360	16·9	4·00	1 190	153	1 350	237	0·88	30·5	0·533	46·0	85·5
	60	406·4	177·8	7·8	12·8	10·2	360·5	6·95	46·2	21 500	1 200	16·8	3·97	1 060	135	1 190	208	0·88	33·9	0·464	32·9	76·0
	54	402·6	177·6	7·6	10·9	10·2	360·5	8·15	47·4	18 600	1 020	16·5	3·85	925	114	1 050	177	0·872	38·5	0·39	22·7	68·4
406×140	46	402·3	142·4	6·9	11·2	10·2	359·7	6·36	52·1	15 600	539	16·3	3·02	778	75·7	888	118	0·87	38·8	0·206	19·2	59·0
	39	397·3	141·8	6·3	8·6	10·2	359·7	8·24	57·1	12 500	411	15·9	2·89	627	58·0	721	91·1	0·859	47·4	0·155	10·6	49·4
356×171	67	364·0	173·2	9·1	15·7	10·2	312·3	5·52	34·3	19 500	1 360	15·1	3·99	1 070	157	1 210	243	0·887	24·4	0·413	55·5	85·4
	57	358·6	172·1	8·0	13·0	10·2	312·3	6·62	39·0	16 100	1 110	14·9	3·92	896	129	1 010	199	0·884	28·9	0·331	33·1	72·2
	51	355·6	171·5	7·3	11·5	10·2	312·3	7·46	42·8	14 200	968	14·8	3·87	796	113	895	174	0·882	32·2	0·286	23·6	64·6
	45	352·0	171·0	6·9	9·7	10·2	312·3	8·81	45·3	12 100	812	14·6	3·78	687	95·0	774	147	0·875	36·9	0·238	15·7	57·0
356×127	39	352·8	126·0	6·5	10·7	10·2	311·2	5·89	47·9	10 100	357	14·3	2·69	572	56·6	654	88·7	0·872	35·3	0·104	14·9	49·4
	33	348·5	125·4	5·9	8·5	10·2	311·2	7·38	52·7	8 200	280	14·0	2·59	471	44·7	540	70·2	0·864	42·2	0·081	8·68	41·8
305×165	54	310·9	7·7	13·7	8·9	265·7	6·09	34·5	11 700	1 060	13·1	3·94	753	127	845	195	0·89	23·7	0·234	34·5	68·4	
	46	307·1	165·7	6·7	11·8	8·9	265·7	7·02	39·7	9 950	897	13·0	3·90	648	108	723	166	0·89	27·2	0·196	22·3	58·9
	40	303·8	165·1	6·1	10·2	8·9	265·7	8·09	43·6	8 520	763	12·9	3·85	561	92·4	624	141	0·888	31·1	0·164	14·7	51·5
305×127	48	310·4	125·2	8·9	14·0	8·9	264·6	4·47	29·7	9 500	460	12·5	2·75	612	73·5	706	116	0·874	23·3	0·101	31·4	60·8
	42	306·6	124·3	8·0	12·1	8·9	264·6	5·14	33·1	8 140	388	12·4	2·70	531	62·5	610	98·2	0·872	26·5	0·0842	21·0	53·2
	37	303·8	123·5	7·2	10·7	8·9	264·6	5·77	36·7	7 160	337	12·3	2·67	472	54·6	540	85·7	0·871	29·6	0·0724	14·9	47·5
305×102	33	312·7	102·4	6·6	10·8	7·6	275·9	4·74	41·8	6 490	193	12·5	2·15	415	37·8	480	59·8	0·886	31·7	0·0041	12·1	41·8
	28	308·9	101·9	6·1	8·9	7·6	275·9	5·72	45·2	5 420	157	112·2	2·08	351	30·8	407	48·9	0·858	37·0	0·0353	7·63	36·3
	25	304·8	101·6	5·8	6·8	7·6	275·9	7·47	47·6	4 390	120	11·8	1·96	288	23·6	338	38·0	0·844	43·8	0·0266	4·65	31·4

Table 2.13 (*cont.*)

Designation		Depth of section	Width of section	Thickness		Root radius	Depth between fillets	Ratios for local buckling		Second moment of area		Radius of gyration		Elastic modulus		Plastic modulus		Buckling parameter	Torsional index	Warping constant	Torsional constant	Area of section
Serial size (mm)	Mass per metre (kg)	D (mm)	B (mm)	Web t (mm)	Flange T (mm)	r (mm)	d (mm)	Flange b/T	Web d/t	XX axis (cm^4)	YY axis (cm^4)	XX axis (cm)	YY axis (cm)	XX axis (cm^3)	YY axis (cm^3)	XX axis (cm^3)	YY axis (cm^3)	u	x	H (dm^6)	J (cm^4)	A (cm^2)
254×146	43	259·6	147·3	7·3	12·7	7·6	218·9	5·80	30·0	6 560	677	10·9	3·51	505	92·0	568	141	0·889	21·1	0·103	24·1	55·1
	37	256·0	146·4	6·4	10·9	7·6	218·9	6·72	34·2	5 560	571	10·8	3·47	434	78·1	485	120	0·889	24·3	0·0858	15·5	47·5
	31	251·5	146·1	6·1	8·6	7·6	218·9	8·49	35·9	4440	449	10·5	3·35	353	61·5	396	94·5	0·879	29·4	0·0662	8·73	40·0
254×102	28	260·4	102·1	6·4	10·0	7·6	225·1	5·10	35·2	4010	178	10·5	2·22	308	34·9	353	54·8	0·873	27·5	0·0279	9·64	36·2
	25	257·0	101·9	6·1	8·4	7·6	225·1	6·07	36·9	3410	148	10·3	2·14	265	29·0	306	45·8	0·864	31·4	0·0228	6·45	32·2
	22	254·0	101·6	5·8	6·8	7·6	225·1	7·47	38·8	2 870	120	10·0	2·05	226	23·6	262	37·5	0·854	35·9	0·0183	4·31	28·4
203×133	30	206·8	133·8	6·3	9·6	7·6	172·3	6·97	27·3	2 890	384	8·72	3·18	279	57·4	313	88·1	0·882	21·5	0·0373	10·2	38·0
	25	203·2	133·4	5·8	7·8	7·6	172·3	8·55	29·7	2 360	310	8·54	3·10	232	46·4	260	71·4	0·876	25·4	0·0295	6·12	32·3
203×102	23	203·2	101·6	5·2	9·3	7·6	169·4	5·46	32·6	2 090	163	8·49	2·37	206	32·1	232	49·5	0·89	22·6	0·0153	6·87	29·0
178×102	19	177·8	101·6	4·7	7·9	7·6	146·8	6·43	31·2	1 360	138	7·49	2·39	153	27·2	171	41·9	0·889	22·6	0·00998	4·37	24·2
152×89	16	152·4	88·9	4·6	7·7	7·6	121·8	5·77	26·5	838	90·4	6·40	2·10	110	20·3	124	31·4	0·889	19·5	0·00473	3·61	20·5
127×76	13	127·0	76·2	4·2	7·6	7·6	96·6	5·01	23·0	477	56·2	5·33	1·83	75·1	14·7	85	22·7	0·893	16·2	0·002	2·92	16·8

Table 2.14 Section properties of Universal Columns to BS4: Part 1

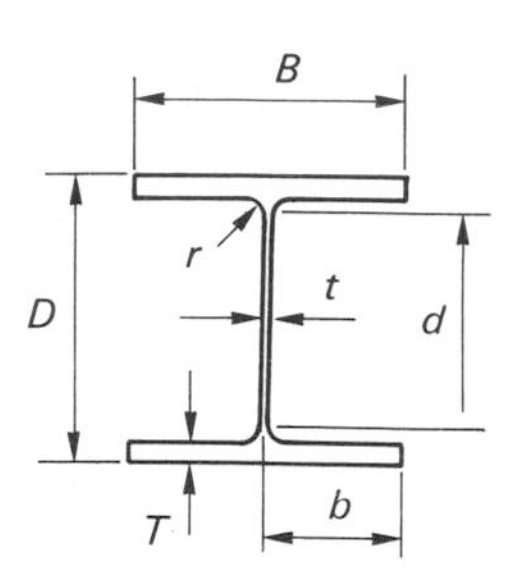

Designation		Depth of section	Width of section	Thickness		Root radius	Depth between fillets	Ratios for local buckling		Second moment of area	
Serial size (mm)	Mass per metre (kg)	D (mm)	B (mm)	Web t (mm)	Flange T (mm)	r (mm)	d (mm)	Flange b/T	Web d/t	XX axis (cm⁴)	YY axis (cm⁴)
356×406	634	474·7	424·1	47·6	77·0	15·2	290·2	2·75	6·10	275 00	98 200
	551	455·7	418·5	42·0	67·5	15·2	290·2	3·10	6·91	227 000	82 700
	467	436·6	412·4	35·9	58·0	15·2	290·2	3·56	8·08	183 000	67 900
	393	491·1	407·0	30·6	49·2	15·2	290·2	4·14	9·48	147 000	55 400
	340	406·4	403·0	26·5	42·9	15·2	290·2	4·70	11·0	122 000	46 800
	287	393·7	399·0	22·6	36·5	15·2	290·2	5·47	12·8	100 000	38 700
	235	381·0	395·0	18·5	30·2	15·2	290·2	6·54	15·7	79 100	31 000
COLCORE	477	427·0	424·4	48·0	53·2	15·2	290·2	3·99	6·05	172 000	68 100
356×368	202	374·7	374·4	16·8	27·0	15·2	290·2	6·93	17·3	66 300	23 600
	177	368·3	372·1	14·5	23·8	15·2	290·2	7·82	20·0	57 200	20 500
	153	362·0	370·2	12·6	20·7	15·2	290·2	8·94	23·0	48 500	17 500
	129	355·6	368·3	10·7	17·5	15·2	290·2	10·5	27·1	40 200	14 600
305×305	283	365·3	321·8	26·9	44·1	15·2	246·6	3·65	9·17	78 800	24 500
	240	352·6	317·9	23·0	37·7	15·2	246·6	4·22	10·7	64 200	20 200
	198	339·9	314·1	19·2	31·4	15·2	246·6	5·00	12·8	50 800	16 200
	158	327·2	310·6	15·7	25·0	15·2	246·6	6·21	15·7	38 700	12 500
	137	320·5	308·7	13·8	21·7	15·2	246·6	7·11	17·9	32 800	10 700
	118	314·5	306·8	11·9	18·7	15·2	246·6	8·20	20·7	27 600	9 010
	97	307·8	304·8	9·9	15·4	15·2	246·6	9·90	24·9	22 200	7 270
254×254	167	289·1	264·5	19·2	31·7	12·7	200·3	4·17	10·4	29 900	9 800
	132	276·4	261·0	15·6	25·3	12·7	200·3	5·16	2·8	22 600	7 520
	107	266·7	258·3	13·0	20·5	12·7	200·3	6·30	15·4	17 500	5 900
	89	260·4	255·9	10·5	17·3	12·7	200·3	7·40	19·1	14 300	4 850
	73	254·0	254·0	8·6	14·2	12·7	200·3	8·94	23·3	11 400	3 870
203×203	86	222·3	208·8	13·0	20·5	10·2	160·9	5·09	12·4	9 460	3 120
	71	215·9	206·2	10·3	17·3	10·2	160·9	5·96	15·6	7 650	2 540
	60	209·6	205·2	9·3	14·2	10·2	160·9	7·23	17·3	6 090	2 040
	52	206·2	203·9	8·0	12·5	10·2	160·9	8·16	20·1	5 260	1 770
	46	203·2	203·2	7·3	11·0	10·2	160·9	9·24	22·0	4 560	1 540
152×152	37	161·8	154·4	8·1	11·5	7·6	123·5	6·71	15·2	2 220	709
	30	157·5	152·9	6·6	9·4	7·6	123·5	8·13	18·7	1 740	558
	23	152·4	152·4	6·1	6·8	7·6	123·5	11·2	20·2	1 260	403

Radius of gyration		*Elastic modulus*		*Plastic modulus*		*Buckling parameter*	*Torsional index*	*Warping constant*	*Torsional constant*	*Area of section*
XX axis (cm)	*YY axis (cm)*	*XX axis* (cm^3)	*YY axis* (cm^3)	*XX axis* (cm^3)	*YY axis* (cm^3)	u	x	H (dm^6)	J (cm^4)	A (cm^2)
18·5	11·0	11 600	4630	14 200	7110	0·843	5·46	38·8	13 700	808
18·0	10·9	9 960	3950	12 100	6060	0·841	6·05	31·1	9 240	702
17·5	10·7	8 390	3290	10 000	5040	0·839	6·86	24·3	5 820	595
17·1	10·5	700	2720	8 230	4160	0·837	7·86	19·0	3 550	501
16·8	10·4	6 030	2320	6 990	3540	0·836	8·85	15·5	2 340	433
16·5	10·3	5 080	1940	5820	2950	0·835	10·2	12·3	1 440	366
16·2	10·2	4 150	1570	4 690	2380	0·834	12·1	9·54	812	300
16·8	10·6	8 080	3210	9 700	4980	0·815	6·91	23·8	5 700	607
16·0	9·57	3 540	1260	3 980	1920	0·844	13·3	7·14	560	258
15·9	9·52	3 100	1100	3 460	1670	0·844	15·0	6·07	383	226
15·8	9·46	2 680	944	2 960	1430	0·844	17·0	5·09	251	195
15·6	9·39	2 260	790	2 480	1200	0·843	19·9	4·16	153	165
14·8	8·25	4 310	1530	5 100	2340	0·855	7·65	6·33	2 030	360
14·5	8·14	3 640	1270	4 250	1950	0·854	8·73	5·01	1 270	306
14·2	8·02	2 990	1030	3 440	1580	0·854	10·2	3·86	734	252
13·9	7·89	2 370	806	2 680	1230	0·852	12·5	2·86	379	201
13·7	7·82	2 050	691	2 300	1050	0·851	14·1	2·38	250	175
13·6	7·75	1 760	587	1 950	892	0·851	16·2	1·97	160	150
13·4	7·68	1 440	477	1 590	723	0·85	19·3	1·55	91.1	150
11·9	6·79	2070	741	2 420	1130	0·852	8·49	1·62	625	123
11·6	6·67	1 630	576	1 870	879	0·85	10·3	1·18	322	169
11·3	6·57	1 310	457	1 490	695	0·848	12·4	0·894	173	137
11·2	6·52	1 100	379	1 230	575	0·849	14·4	0·716	104	114
11·1	6·46	894	305	989	462	0·849	17·3	0·557	57·3	92·9
9·27	5·32	851	299	979	456	0·85	10·2	0·317	138	110
9·16	5·28	708	246	802	374	0·852	11·9	0·25	81·5	91·1
8·96	5·19	581	199	652	303	0·847	14·1	0·195	46·6	75·8
8·90	5·16	510	174	568	264	0·848	15·8	0·166	32·0	66·4
8·81	5·11	449	151	497	230	0·846	17·7	0·142	22·2	58·8
6·84	3·87	274	91·8	310	140	0·848	13·3	0·04	19·5	47·4
6·75	3·82	221	73·1	247	111	0·848	16·0	0·0306	10·5	38·2
6·51	3·68	166	52·9	184	80·9	0·837	20·4	0·0214	4·87	29·8

Table 2.15 Section properties of joists to BS4: Part 1

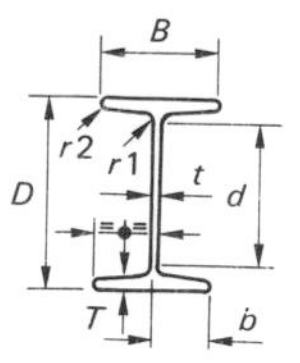

Designation		Depth of section	Width of section	Thickness		Root radius	Toe radius	Depth between fillets	Ratios for local buckling		Second moment of area	
Serial size (mm)	Mass per metre (kg)	D (mm)	B (mm)	Web t (mm)	Flange T (mm)	r1 (mm)	r2 (mm)	d (mm)	Flange b/T	Web d/t	XX axis (cm⁴)	YY axis (cm⁴)
254×203	81·85	254·0	203·2	10·2	19·9	19·6	9·7	166·7	5·11	16·3	12 000	2280
254×114	37·20	254·0	114·3	7·6	12·8	12·4	6·1	199·2	4·46	26·2	5 090	270
203×152	52·09	203·2	152·4	8·9	16·5	15·5	7·6	133·3	4·62	15·0	4 790	813
152×127	37·20	152·4	127·0	10·4	13·2	13·5	6·6	94·4	4·81	9·08	1 820	379
127×114	29·76	127·0	114·3	10·2	11·5	9·9	4·8	79·4	4·97	7·78	979	242
	26·79	127·0	114·3	7·4	11·4	9·9	5·0	79·5	5·01	10·7	945	235
127×76	16·37	127·0	76·2	5·6	9·6	9·4	4·6	86·6	3·97	15·5	569	60·5
114×114	26·79	114·3	114·3	9·5	10·7	14·2	3·2	60·9	5·34	6·41	735	223
102×102	23·07	101·6	101·6	9·5	10·3	11·1	3·2	55·2	4·93	5·81	486	154
102×44	7·44[a]	101·6	44·5	4·3	6·1	6·9	3·3	74·7	3·65	17·4	153	7·74
89×89	19·35	88·9	88·9	9·5	9·9	11·1	3·2	44·2	4·49	4·65	307	101
76×76	14·67[a]	76·2	80·0	8·9	8·4	9·4	4·6	38·1	4·76	4·28	172	60·8
	12·65	76·2	76·2	5·1	8·4	9·4	4·6	38·0	4·54	7·45	158	52·0

[a]These sections are only rolled to specific order

Table 2.16 Section properties of universal bearing piles to BS4: Part 1

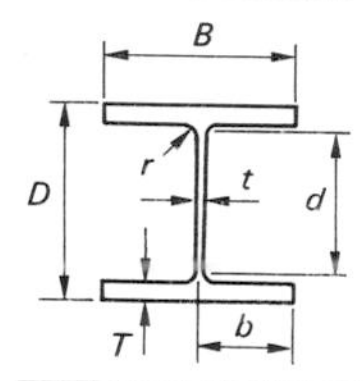

Designation		Depth of section	Width of section	Thickness		Root radius	Depth between fillets	Ratios for local buckling		Second moment of area	
Serial size (mm)	Mass per metre (kg)	D (mm)	B (mm)	Web t (mm)	Flange T (mm)	r (mm)	d (mm)	Flange b/T	Web d/t	XX axis (cm⁴)	YY axis (cm⁴)
356×368	174	361·5	378·1	20·4	20·4	15·2	290·2	9·27	14·2	51 100	18 400
	152	356·4	375·5	17·9	17·9	15·2	290·2	10·5	16·2	43 900	15 800
	153	351·9	373·3	15·6	15·6	15·2	290·2	12·0	18·6	37 800	13 600
	109	346·4	370·5	12·9	12·9	15·2	290·2	14·4	22·5	30 500	10 900
305×305	223	338·0	325·4	30·5	30·5	15·2	246·6	5·33	8·09	52 800	17 600
	186	328·3	320·5	25·6	25·6	15·2	246·6	6·26	9·63	42 600	14 100
	149	318·5	315·6	20·7	20·7	15·2	246·6	7·62	11·9	33 000	10 900
	126	312·6	312·7	17·8	17·8	15·2	246·6	8·78	13·9	27 700	9 070
	110	307·9	310·3	15·4	15·4	15·2	246·6	10·1	16·0	23 600	7 690
	95	303·8	308·3	13·4	13·4	15·2	246·6	11·5	18·4	20 100	6 530
	88	301·7	307·2	12·3	12·3	15·2	246·6	12·5	20·0	18 400	5 960
	79	299·2	306·0	11·1	11·1	15·2	246·6	13·8	22·2	16 400	5 290
254×254	85	254·3	259·7	14·3	14·3	12·7	200·3	9·08	14·0	12 300	4 190
	71	249·9	257·5	12·1	12·1	12·7	200·3	10·6	16·6	10 200	3 450
	63	246·9	256·0	10·6	10·6	12·7	200·3	12·1	18·9	8 970	2 970
203×203	54	203·9	207·2	11·3	11·3	10·2	160·9	9·17	14·2	4 990	1 680
	45	200·2	205·4	9·5	9·5	10·2	160·9	10·8	16·9	4 080	1 370

Radius of gyration		*Elastic modulus*		*Plastic modulus*		*Buckling parameter*	*Torsional index*	*Warping constant*	*Torsional constant*	*Area of section*
XX axis (cm)	*YY axis (cm)*	*XX axis (cm^3)*	*YY axis (cm^3)*	*XX axis (cm^3)*	*YY axis (cm^3)*	*u*	*x*	*H (dm^6)*	*J (cm^4)*	*A (cm^2)*
10·7	4·67	946	224	1080	370	0·89	11·0	0·312	153	104
10·4	2·39	401	47·2	460	79·3	0·885	18·6	0·0393	25·5	47·4
8·49	3·50	471	107	540	176	0·891	10·7	0·0709	64·9	66·4
6·19	2·82	239	59·6	279	99·8	0·866	9·29	0·0183	34·2	47·5
5·12	2·55	154	42·3	181	70·8	0·853	8·74	0·00807	20·9	37·2
5·26	2·63	149	41·2	172	68·1	0·869	9·30	0·00787	16·9	34·1
5·21	1·70	89·7	15·9	104	26·3	0·891	11·8	0·00209	6·69	21·0
4·62	2·55	129	39·1	151	65·6	0·841	7·90	0·00599	19·0	34·4
4·07	2·29	95·7	30·4	113	50·7	0·836	7·39	0·00321	14·4	29·4
4·01	0·904	30·0	3·48	35·3	5·99	0·871	14·9	0·000177	1·25	9·48
3·51	2·02	69·0	22·8	82·8	38·0	0·83	6·54	0·00158	11·6	24·9
3·00	1·78	45·1	15·2	54·2	25·7	0·82	6·42	0·000699	6·83	19·1
3·12	1·79	41·6	13·7	48·8	22·5	0·852	7·16	0·000597	4·67	16·3

Radius of gyration		*Elastic modulus*		*Plastic modulus*		*Buckling parameter*	*Torsional index*	*Warping constant*	*Torsional constant*	*Area of section*
XX axis (cm)	*YY axis (cm)*	*XX axis (cm^3)*	*YY axis (cm^3)*	*XX axis (cm^3)*	*YY axis (cm^3)*	*u*	*x*	*H (dm^6)*	*J (cm^4)*	*A (cm^2)*
15·2	9·11	2830	976	3190	1500	0·821	15·7	5·36	334	222
15·1	9·03	2460	841	2760	1290	0·821	17·8	4·53	224	194
15·0	8·96	2150	727	2400	1110	0·822	20·2	3·84	151	169
14·8	8·87	1760	588	1950	897	0·823	24·2	3·03	84·3	138
13·6	7·85	3130	1080	3660	1680	0·826	9·51	4·15	953	285
13/4	7·71	2600	880	3010	1370	0·827	11·1	3·23	564	237
13·2	7·56	2080	689	2370	1060	0·828	13·5	2·41	297	190
13·1	7·47	1770	580	2010	893	0·829	15·5	1·97	188	162
13·0	7·40	1530	496	1720	761	0·83	17·7	1·64	123	140
12·9	7·33	1320	424	1480	649	0·831	20·1	1·38	81·0	121
12·8	7·30	1220	388	1360	594	0·831	21·7	1·25	63·9	112
12·8	7·26	1100	346	1220	529	0·832	23·8	1·10	47·0	100·0
10·7	6·22	965	323	1090	496	0·826	15·6	0·603	81·6	108
10·6	6·15	813	268	911	411	0·827	18·2	0·488	49·7	91·1
10·5	6·11	711	232	793	355	0·827	20·5	0·415	33·8	79·7
8·54	4·96	489	162	553	250	0·827	15·9	0·156	32·3	68·4
8·46	4·90	408	133	457	204	0·828	18·7	0·124	19·0	57·0

Table 2.17 Section properties of channels to BS4: Part 1

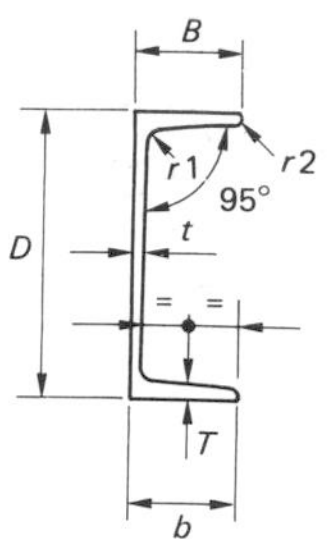

Designation		*Depth of section*	*Width of section*	*Thickness*		*Root radius*	*Toe radius*	*Depth between fillets*	*Ratios for local buckling*		*Second moment of area*	
Serial size (mm)	*Mass per metre (kg)*	*D (mm)*	*B (mm)*	*Web t (mm)*	*Flange T (mm)*	*r1 (mm)*	*r2 (mm)*	*d (mm)*	*Flange b/T*	*Web d/t*	*XX axis* (cm^4)	*YY axis* (cm^4)
432×102	65·54	431·8	101·6	12·2	16·8	15·2	4·8	362·4	6·05	29·7	21 400	629
381×102	55·10	381·0	101·6	10·4	16·3	15·2	4·8	312·4	6·23	30·0	14 900	580
305×102	46·18	304·8	101·6	10·2	14·8	15·2	4·8	239·2	6·86	23·5	8 210	499
305×89	41·69	304·8	88·9	10·2	13·7	13·7	3·2	245·4	6·49	24·1	7 060	325
254×89	35·74	254·0	88·9	9·1	13·6	13·7	3·2	194·7	6·54	21·4	4 450	302
254×76	28·29	254·0	76·2	8·1	10·9	12·2	3·2	203·8	6·99	25·2	3 370	163
229×89	32·76	228·6	88·9	8·6	13·3	13·7	3·2	169·8	6·68	19·7	3 390	285
229×76	26·06	228·6	76·2	7·6	11·2	12·2	3·2	178·0	6·80	23·4	2 610	159
203×89	29·78	203·2	88·9	8·1	12·9	13·7	3·2	145·2	6·89	17·9	2 490	264
203×76	23·82	203·2	76·2	7·1	11·2	12·2	3·2	152·5	6·80	21·5	1 950	151
178×89	26·81	177·8	88·9	7·6	12·3	137	3·2	121·0	7·23	15·9	1 750	241
178×76	20·84	177·8	76·2	6·6	10·3	12·2	3·2	128·8	7·40	19·5	1 340	134
152×89	23·84	152·4	88·9	7·1	11·6	13·7	3·2	97·0	7·66	13·7	1 170	215
152×76	17·88	152·4	76·2	6·4	9·0	12·2	2·4	105·9	8·47	16·5	852	114
127×64	14·90	127·0	63·5	6·4	9·2	10·7	2·4	84·0	6·90	13·1	483	67·2
102×51	10·42	101·6	50·8	6·1	7·6	9·1	2·4	65·7	6·68	10·8	208	29·1
76×38	6·70	76·2	38·1	5·1	6·8	7·6	2·4	45·8	5·60	8·98	74·1	10·7

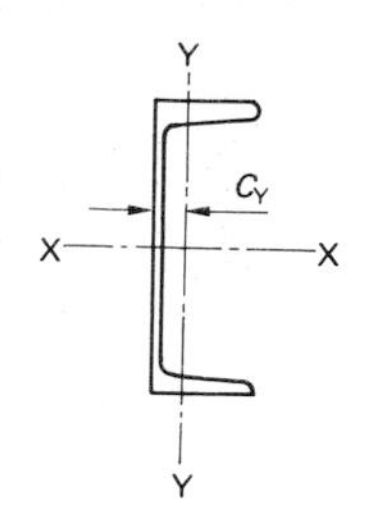

Radius of gyration		Elastic modulus		Plastic modulus		Buckling parameter	Torsional index	Warping constant	Torsional constant	Area of section
XX axis (cm)	*YY axis (cm)*	*XX axis (cm^3)*	*YY axis (cm^3)*	*XX axis (cm^3)*	*YY axis (cm^3)*	*u*	*x*	*H (dm^6)*	*J (cm^4)*	*A (cm^2)*
16·0	2·74	991	80·1	1210	153	0·876	24·6	0·217	61·0	83·5
14·6	2·87	782	75·9	933	144	0·895	22·7	0·153	46·0	70·2
11·8	2·91	539	66·6	638	128	0·9	18·9	0·0842	35·4	58·8
11·5	2·48	463	48·5	557	92·6	0·887	20·4	0·0551	27·6	53·1
9·88	2·58	350	46·7	414	89·6	0·906	17·1	0·0347	22·9	45·5
9·67	2·12	265	28·2	317	54·1	0·886	21·2	0·0194	12·3	36·0
9·01	2·61	296	44·8	348	86·4	0·912	15·5	0·0263	20·4	41·7
8·87	2·19	228	28·2	270	54·2	0·9	18·8	0·0151	11·4	33·2
8·10	2·64	245	42·3	287	81·6	0·915	14·1	0·0192	17·8	37·9
8·02	2·23	192	27·6	225	53·3	0·911	16·7	0·0112	10·4	30·3
7·16	2·66	197	39·3	230	75·4	0·915	12·7	0·0134	15·1	34·2
7·10	2·25	150	24·7	175	48·1	0·911	15·5	0·00764	8·13	26·5
6·20	2·66	153	35·7	178	68·1	0·909	11·3	0·00881	12·4	30·4
6·11	2·24	112	21·0	130	41·3	0·902	14·5	0·00486	5·94	22·8
5·04	1·88	76·0	15·3	89·4	29·3	0·91	11·7	0·00187	4·92	19·0
3·96	1·48	40·9	8·160	48·8	15·7	0·9	10·8	0·000513	2·55	13·3
2·95	1·12	19·5	4·070	23·4	7·76	0·908	9·170	0·000101	1·23	8·53

Table 2.18 Section properties of equal angles to BS 4848: Part 4

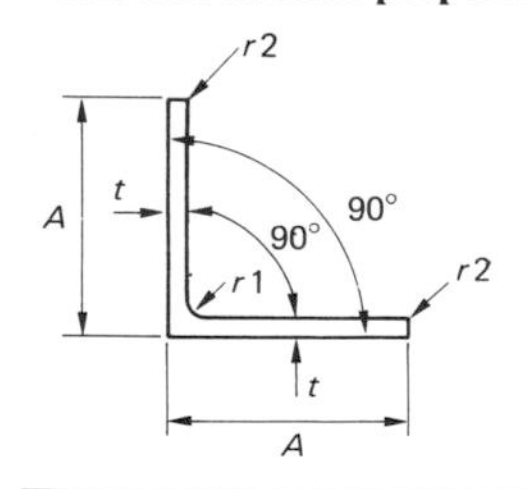

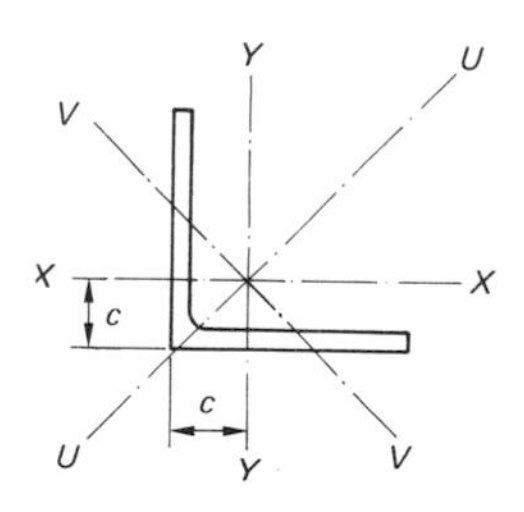

Designation		Mass per metre	Radius		Area of section	Distance of centre of gravity	Second moment of area			Radius of gyration			Elastic modulus
Size A A (mm)	*Thickness t*	*(kg)*	*Root r1 (mm)*	*Toe r2 (mm)*	*(cm^2)*	*C_x and C_y (cm)*	*XX, YY axis (cm^4)*	*UU axis (cm^4)*	*VV axis (cm^4)*	*XX, YY axis (cm)*	*UU axis (cm)*	*VV axis (cm)*	*XX, YY axis (cm^3)*
250×250	35	128	20·0	4·8	150	7·49	9250	14 600	3860	7·53	9·47	4·86	529
	32	118	20·0	4·8	150	7·38	8600	13 600	3560	7·57	9·53	4·87	488
	28	104	20·0	4·8	133	7·23	7690	12 200	3170	7·61	9·59	4·89	433
	25	93·6	20·0	4·8	119	7·12	6970	11 100	2860	7·65	9·64	4·9	390
200×200	24	71·1	18·0	4·8	90·6	5·84	3330	5 280	1380	6·06	7·64	3·9	235
	20	59·9	18·0	4·8	76·3	5·68	2850	4 530	1170	6·11	7·7	3·92	199
	18	54·2	18·0	4·8	69·1	5·6	2600	4 130	1070	6·13	7·73	3·93	181
	16	48·5	18·0	4·8	61·8	5·52	2340	3 720	960	6·16	7·76	3·94	162

Table 2.18 (*cont.*)

Designation		*Mass per metre (kg)*	*Radius*		*Area of section (cm^2)*	*Distance of centre of gravity C_x and C_y (cm)*	*Second moment of area*			*Radius of gyration*			*Elastic modulus XX, YY axis (cm^3)*
Size A A (mm)	*Thickness t*		*Root r1 (mm)*	*Toe r2 (mm)*			*XX, YY axis (cm^4)*	*UU axis (cm^4)*	*VV axis (cm^4)*	*XX, YY axis (cm)*	*UU axis (cm)*	*VV axis (cm)*	
150×150	18	40·1	16·0	4·8	51·0	4·37	1050	1 660	435	4·54	5·71	2·92	98·7
	15	33·8	16·0	4·8	43·0	4·25	898	1 430	370	4·57	5·76	2·93	83·5
	12	27·3	16·0	4·8	34·8	4·12	737	1 170	303	4·6	5·8	2·95	67·7
	10	23·0	16·0	4·8	29·3	4·03	624	991	258	4·62	5·82	2·97	56·9
120×120	15	26·6	13·0	4·8	33·9	3·51	445	705	185	3·62	4·56	2·33	52·4
	12	21·6	13·0	4·8	27·5	3·4	368	584	152	3·65	4·6	2·35	42·7
	10	18·2	13·0	4·8	23·2	3·13	313	497	129	3·67	4·63	2·36	36·0
	8	14·7	13·0	4·8	18·7	3·23	255	405	105	3·69	4·65	2·37	29·1
100×100	15	21·9	12·0	4·8	27·9	3·02	249	393	104	2·98	3·75	1·93	35·6
	12	17·8	12·0	4·8	22·7	2·9	207	328	85·7	3·02	3·8	1·94	29·1
	8	12·2	12·0	4·8	15·5	2·74	145	230	59·9	3·06	3·85	1·96	19·9
90×90	12	15·9	11·0	4·8	20·3	2·66	148	234	61·7	2·7	3·4	1·74	23·3
	10	13·4	11·0	4·8	17·1	2·58	127	201	52·6	2·72	3·43	1·75	19·8
	8	10·9	11·0	4·8	13·9	2·5	104	166	43·1	2·74	3·45	1·76	16·1
	7	9·61	11·0	4·8	12·2	2·45	92·5	147	38·3	2·75	3·46	1·77	14·1
	6	8·3	11·0	4·8	10·6	2·41	80·3	127	33·3	2·76	3·47	1·78	12·2
80×80	10	11·9	10·0	4·8	15·1	2·34	87·5	139	36·4	2·41	3·03	1·55	15·4
	8	9·63	10·0	4·8	12·3	2·26	72·2	115	29·9	2·43	3·06	1·56	12·6
	6	7·34	10·0	4·8	9·35	2·17	55·8	88·5	23·1	2·44	3·08	1·57	9·57
70×70	10	10·3	9·0	2·4	13·1	2·09	57·2	90·5	24·0	2·09	2·63	1·35	11·7
	8	8·36	9·0	2·4	10·6	2·01	47·5	75·3	19·7	2·11	2·66	1·36	9·52
	6	6·38	9·0	2·4	8·13	1·93	36·9	58·5	15·3	2·13	2·68	1·37	7·27
60×60	10	8·69	8·0	2·4	11·1	1·85	34·9	55·1	14·8	1·78	2·23	1·16	8·41
	8	7·09	8·0	2·4	9·03	1·77	29·2	46·1	12·2	1·8	2·26	1·16	6·89
	6	5·42	8·0	2·4	6·91	1·69	22·8	36·1	9·44	1·82	2·29	1·17	5·29
	5	4·57	8·0	2·4	5·82	1·64	19·4	30·7	8·03	1·82	2·3	1·17	4·45
50×50	8	5·82	7·0	2·4	7·41	1·52	16·3	25·7	6·87	1·48	1·86	0·963	4·68
	6	4·47	7·0	2·4	5·69	1·45	12·8	20·3	5·34	1·5	1·89	0·968	3·61
	5	3·77	7·0	2·4	4·8	1·4	11·0	17·4	4·55	1·51	1·9	0·973	3·05
	4	3·06	7·0	2·4	3·89	1·36	8·97	14·2	3·73	1·52	1·91	0·979	2·46
	3	2·33	7·0	2·4	2·96	1·31	6·86	10·8	2·88	1·52	1·91	0·986	1·86
45×45	6	4·0	7·0	2·4	5·09	1·32	9·16	14·5	3·83	1·34	1·69	0·867	2·88
	5	3·38	7·0	2·4	4·3	1·28	7·84	12·4	3·26	1·35	1·7	0·871	2·43
	4	2·74	7·0	2·4	3·49	1·23	6·43	10·2	2·68	1·36	1·71	0·876	1·97
	3	2·09	7·0	2·4	2·66	1·18	4·93	7·78	2·07	1·36	1·71	0·882	1·49
40×40	6	3·52	6·0	2·4	4·48	1·2	6·31	9·98	2·65	1·19	1·49	0·77	2·26
	5	2·97	6·0	2·4	3·79	1·16	5·43	8·59	2·26	1·2	1·51	0·773	1·91
	4	2·42	6·0	2·4	3·08	1·12	4·47	7·09	1·86	1·21	1·52	0·777	1·55
	3	1·84	6·0	2·4	2·35	1·07	3·45	5·45	1·44	1·21	1·52	0·783	1·18
30×30	5	2·18	5·0	2·4	2·78	0·918	2·16	3·41	0·917	0·883	1·11	0·575	1·04
	4	1·78	5·0	2·4	2·27	0·878	1·8	2·85	0·754	0·892	1·12	0·577	0·85
	3	1·36	5·0	2·4	1·74	0·835	1·4	2·22	0·585	0·899	1·13	0·581	0·649
25×25	5	1·77	3·5	2·4	2·26	0·799	1·21	1·9	0·524	0·731	0·915	0·481	0·711
	4	1·45	3·5	2·4	1·85	0·762	1·02	1·61	0·43	0·741	0·931	0·482	0·586
	3	1·11	3·5	2·4	1·42	0·723	0·803	1·27	0·334	0·751	0·945	0·484	0·452

Note: A 100 × 100 × 10 mm angle is also frequently rolled; as an ISO size its properties are given in Appendix A (Table A1) to BS 4848: Part 4. Other non-standard sections, particularly other thicknesses of the standard range, may also be available. Enquiries should be made to BSC General Steels Group BSC Sections.

T-sections: Ts cut from Unviersal Beams and Columns listed in this publication are available in all sizes. Enquiries should be made to BSC General Steels Group BSC Sections. Properties are listed in Volume 1 of the *Steelwork Design Guide to BS 5950: Part 1* a publication produced by the Steel Construction Institute.

Bulb flats: These are now produced in metric sizes and are listed in BS 4848: Part 5. Enquiries should be made to BSC General Steels Group BSC Sections.

Table 2.19 Section properties of unequal angles to BS 4848: Part 4

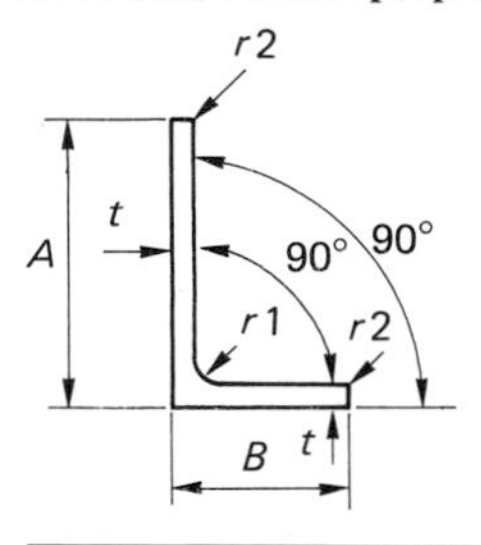

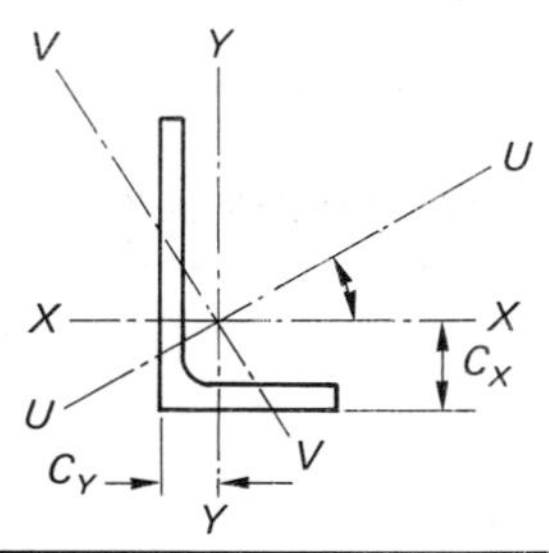

Designation: Size AB (mm)	Designation: Thickness t (mm)	Mass per metre (kg)	Radius: Root r1 (mm)	Radius: Toe r2 (mm)	Area of section (cm²)	Distance centre of gravity C_X (cm)	Distance centre of gravity C_Y (cm)	Second moment of area: XX axis (cm⁴)	Second moment of area: YY axis (cm⁴)	Second moment of area: UU axis (cm⁴)	Second moment of area: VV axis (cm⁴)	Radius of gyration: XX axis (cm)	Radius of gyration: YY axis (cm)	Radius of gyration: UU axis (cm)	Radius of gyration: VV axis (cm)	Elastic modulus: XX axis (cm³)	Elastic modulus: YY axis (cm³)	Angle XX axis to UU axis (tan)
200×150	18	47·1	15·0	4·8	60·0	6·33	3·85	2 380	1 150	2 900	618	6·29	4·37	6·96	3·21	174	103	0·548
	15	39·6	15·0	4·8	50·5	6·21	3·73	2 020	979	2 480	526	6·33	4·4	7·0	3·23	147	86·9	0·551
	12	32·0	15·0	4·8	40·8	6·08	3·61	1 650	803	2 030	430	6·36	4·44	7·04	3·25	119	70·5	0·552
200×100	15	33·7	15·0	4·8	43·0	7·16	2·22	1 760	299	1 860	194	6·4	2·64	6·58	2·12	137	38·4	0·26
	12	27·3	15·0	4·8	34·8	7·03	2·1	1 440	247	1 530	159	6·43	2·67	6·63	2·14	111	31·3	0·262
	10	23·0	15·0	4·8	29·2	6·93	2·01	1 220	210	1 290	135	6·46	2·68	6·65	2·15	93·2	26·3	0·263
150×90	15	26·6	12·0	4·8	33·9	5·21	2·23	761	205	841	126	4·74	2·46	4·98	1·93	77·7	30·4	0·354
	12	21·6	12·0	4·8	27·5	5·08	2·12	627	171	694	104	4·77	2·49	5·02	1·94	63·3	24·8	0·358
	10	18·2	12·0	4·8	23·2	5·0	2·04	533	146	591	88·3	4·8	2·51	5·05	1·95	53·3	21·0	0·36
150×75	15	24·8	11·0	4·8	31·6	5·53	1·81	713	120	754	78·8	4·75	1·94	4·88	1·58	75·3	21·0	0·254
	12	20·1	11·0	4·8	25·7	5·41	1·69	589	99·9	624	64·9	4·79	1·97	4·93	1·59	61·4	17·2	0·259
	10	17·0	11·0	4·8	21·6	5·32	1·61	501	85·8	532	55·3	4·81	1·99	4·96	1·6	51·8	14·6	0·261
125×75	12	17·8	11·0	4·8	22·7	4·31	1·84	354	95·5	391	58·5	3·95	2·05	4·15	1·61	43·2	16·9	0·354
	10	15·0	11·0	4·8	19·1	4·23	1·76	302	82·1	334	49·9	3·87	2·07	4·18	1·61	36·5	14·3	0·357
	8	12·2	11·0	4·8	15·5	4·14	1·68	247	67·6	274	40·9	4·0	2·09	4·21	1·63	29·6	11·6	0·36
75	12	15·4	10·0	4·8	19·7	3·27	2·03	189	90·2	230	49·5	3·1	2·14	3·42	1·59	28·0	16·5	0·54
	10	13·0	10·0	4·8	16·6	3·19	1·95	162	77·6	197	42·2	3·12	2·16	3·45	1·59	23·8	14·0	0·544
	8	10·6	10·0	4·8	13·5	3·1	1·87	133	64·1	162	34·6	3·14	2·18	3·47	1·6	19·3	11·4	0·547
100×65	10	12·3	10·0	4·8	15·6	3·36	1·63	154	51·0	175	30·1	3·14	1·81	3·35	1·39	23·2	10·5	0·41
	8	9·94	10·0	4·8	12·7	3·27	1·55	127	42·2	144	24·8	3·16	1·83	3·37	1·4	18·9	8·54	0·413
	7	8·77	10·0	4·8	11·2	3·23	1·51	113	37·6	128	22·0	3·17	1·83	3·39	1·4	16·6	7·53	0·415
80×60	8	8·34	8·0	4·8	10·6	2·55	1·56	66·3	31·8	80·0	17·3	2·5	1·73	2·76	1·27	12·2	7·16	0·544
	7	7·36	8·0	4·8	9·38	2·51	1·52	59·0	28·4	72·0	15·4	2·51	1·74	2·77	1·28	10·7	6·34	0·546
	6	6·37	8·0	4·8	8·11	2·47	1·48	51·4	24·8	62·8	13·4	2·52	1·75	2·78	1·29	9·2	5·49	0·547
75×50	8	7·39	7·0	2·4	9·41	2·52	1·29	52·0	18·4	59·6	10·8	2·35	1·4	2·52	1·07	10·4	4·95	0·43
	6	5·65	7·0	2·4	7·19	2·44	1·21	40·5	14·4	46·6	8·36	2·37	1·42	2·55	1·08	8·0	3·81	0·435
65×50	8	6·75	6·0	2·4	8·6	2·11	1·37	34·8	17·7	43·0	9·57	2·01	1·44	2·23	1·05	7·9	4·89	0·569
	6	5·16	6·0	2·4	6·58	2·04	1·29	27·2	14·0	33·8	7·43	2·03	1·46	2·27	1·06	6·1	3·77	0·575
	5	4·35	6·0	2·4	5·54	1·99	1·25	23·2	11·9	28·8	6·32	2·05	1·47	2·28	1·07	5·1	3·19	0·577
60×30	6	3·99	6·0	2·4	5·08	2·2	0·72	18·2	3·0	19·2	1·99	1·89	0·771	1·95	0·626	4·7	1·32	0·252
	5	3·37	6·0	2·4	4·29	2·15	0·68	15·6	2·6	16·5	1·7	1·9	0·779	1·96	0·629	4·0	1·12	0·256
40×25	4	1·93	4·0	2·4	2·46	1·36	0·62	3·8	1·1	4·3	0·70	1·26	0·688	1·33	0·534	1·4	0·61	0·38

Note: Additional non-standard sizes may be available, especially other thicknesses of the standard range and certain sizes in the old imperial range, namely 125 × 75 × 6·5 and 137 102 × 9·5, 7·9 and 6·4 (purlin angles) and 100 × 75 × 6·5. Enquiries should be made to BSC General Steels Group BSC Sections.

3

Structural timber

The relevant British Standard is BS 5268, *Structural Use of Timber*, Part 2: 1988, Code of practice for permissible stress design, materials and workmanship. This chapter deals with the structural design of flexural and compression solid timber members in accordance with the procedures described in Section 2 of BS 5268: Part 2.

It is intended to publish BS 5268 in seven parts with Part 2 replacing CP112: Part 2: 1971, which has now been withdrawn. The full list is as follows:

Part 1: Limit state design, materials and workmanship.
Part 2: Permissible stress design, materials and workmanship.
Part 3: Code of practice for trussed rafter roofs.
Part 4: Fire resistance of timber structures.
Part 5: Preservation treatments for constructional timber.
Part 6: Code of practice for timber frame wall design.
Part 7: Recommendations for the calculation basis for span tables.

Parts 1 and 7 are still in the process of preparation.

3.1 Design requirements

In order to allow a greater flexibility of choice of timber suitable for structural applications the code has introduced a method of classifying each species of hardwood and softwood under one of nine strength classes. Table 3.1 gives the grade stresses and moduli of elasticity for the nine strength classes and Table 3.2 relates the commoner softwood species and stress grade combinations with their relevant strength class. The advantage to both the designer and the client is that by specifying the timber by strength class, rather than by the more specific method of species and grade, the supplier is able to supply a species of timber on the basis of cost and availability for a particular strength class.

Stress-graded timber

The grading system used in Table 3.2 is derived from BS 4987: 1973, *Specification for Timber Grades for Structural Use*. BS 4987 defines two visual and four machine grades. Visual stress grading is based on certain physical characteristics of the timber (for example, the number and size of knots and the slope of the grain). The two grades are:

1. General structural (GS), and
2. Special structural (SS).

Mechanical stress grading is based on a non-destructive test that assesses the modulus of elasticity of the timber, and the four grades are:

1. General structural (MGS);
2. Special structural (MSS);
3. M50; and
4. M75.

Table 3.1 Softwood species/grade[a] combinations which satisfy the requirement for strength classes: graded to BS 4978

Standard name	*Strength class*				
	SC1	*SC2*	*SC3*	*SC4*	*SC5*
Imported					
Parana pine			GS	SS	
Pitch pine (Caribbean)			GS		SS
Redwood			GS/M50	SS	M75
Whitewood			GS/M50	SS	M75
Western red cedar	GS	SS			
Douglas fir-larch (Canada)			GS	SS	
Douglas fir-larch (USA)			GS	SS	
Hem-fir (Canada)			GS/M50	SS	M75
Hem-fir (USA)			GS	SS	
Spruce-pine-fir (Canada)			GS/M50	SS/M75	
Western whitewoods (USA)	GS		SS		
Southern pine (USA)			GS	SS	
British grown					
Douglas fir		GS	M50/SS		M75
Larch			GS	SS	
Scots pine			GS/M50	SS	M75
Corsican pine		GS	M50	SS	M75
European spruce	GS	M50/SS	M75		
Sitka spruce	GS	M50/SS	M75		

[a] Machine grades MGS and MSS are interchangeable with GS and SS grades, respectively.
The S6, S8, MS6 and MS8 grades of the ECE *Recommended Standard for Stress Grading of Coniferous Sawn Timber* (1982) may be substituted for GS, SS, MGS and MSS, respectively.

Table 3.2 Grade stress and moduli of elasticity for strength classes: for the dry exposure condition

Strength class	*Bending parallel to grain* (N/mm^2)	*Tension parallel to grain* (N/mm^2)	*Compression parallel to grain* (N/mm^2)	*Compression perpendicular to grain[a]* (N/mm^2)	(N/mm^2)	*Shear parallel to grain* (N/mm^2)	*Modulus of elasticity* *Mean* (N/mm^2)	*Minimum* (N/mm^2)	*Approximate density[b]* (kg/m^3)
SC1	2·8	2·2[c]	3·5	2·1	1·2	0·46	6800	4500	540
SC2	4·1	2·5[c]	5·3	2·1	1·6	0·66	8000	5000	540
SC3	5·3	3·2[c]	6·8	2·2	1·7	0·67	8800	5800	540
SC4	7·5	4·5[c]	7·9	2·4	1·9	0·71	9900	6600	590
SC5	10·0	6·0[c]	8·7	2·8	2·4	1·00	10700	7100	590/760
SC6[d]	12·5	7·5	12·5	3·8	2·8	1·50	14100	11800	840
SC7[d]	15·0	9·0	14·5	4·4	3·3	1·75	16200	13600	960
SC8[d]	17·5	10·5	16·5	5·2	3·9	2·00	18700	15600	1080
SC9[d]	20·5	12·3	19·5	6·1	4·6	2·25	21600	18000	1200

[a] When the specification specifically prohibits wane at bearing areas, the higher values of compression perpendicular to the grain stress may be used, otherwise the lower values apply.
[b] Since many species may contribute to any of the strength classes, the values of density given in this table may be considered only crude approximations. When a more accurate value is required it may be necessary to identify individual species. The higher value for SC5 is more appropriate for hardwoods.
[c] Note that the Light Framing, Stud, Structural Light Framing No.3 and Joist and Plank No. 3 grades should not be used in tension.
[d] Classes SC6, SC7, SC8 and SC9 will usually comprise the denser hardwoods.

Moisture content

The moisture content of timber, measured as a percentage of its oven-dried weight, is an important factor both from the point of view of its structural capacity and of its durability. Section 1 (5.4) draws to the attention of the designer, among other points, the following

1. At a moisture content below about 30% timber

shrinks or swells, and its strength properties increase or decrease according to its moisture content.
2. Wood is less prone to decay if its moisture content is below 25% and may be considered immune below 20%.
3. All timber, whether imported or British grown, which is thoroughly air dried in the UK normally attains a moisture content between 17% and 23%, depending upon weather conditions at the end of the drying period.
4. Ideally, timber should have a moisture content, when installed, close to the content it will attain in service. Table 3.3 gives average values for softwoods and hardwoods for selected categories of end use. However, this should only be seen as a general guide.
5. Care should be taken on site to ensure that material supplied in a dry condition is adequately protected from the weather.

Service exposure conditions

Because of the effect of moisture content on strength properties the permissible stresses used in design should be those corresponding to the moisture content that the particular member will attain in service. The code recognizes that it is not possible to cover all service conditions but defines the following two exposure conditions:

1. *Dry exposure*. All service conditions where the air temperature and humidity would result in solid timber attaining an equilibrium moisture content not exceeding 18% for any significant period. This condition would include most covered buildings and internal uses. Such stresses are identified as 'dry stresses' (see Table 3.1).
2. *Wet exposure*. All service conditions, either in contact with water or where the air temperature and humidity would result in solid timber attaining an equilibrium moisture content exceeding 18% for any significant period. The permissible stress values given in Table 3.1 should be multiplied by the factor K_2 (see Table 3.5) to obtain the appropriate wet stress values.

Symbols and subscripts

The symbols used in this chapter are as listed in Section 1, Part 3 of the code, but will also be identified when used in the text. However, the code also uses subscripts with the symbols which, for the sake of space, will only be identified here. The subscripts used are:

1. Type of force, stress, etc:
   - c: compression
   - m: bending
   - t: tension
   - s: shear
2. Significance:
   - a: applied
   - adm: permissible
   - e: effective
   - mean: arithmetic mean
3. Geometry:
   - ∥: parallel (to the grain)
   - ⊥: perpendicular (to the grain)

As an example, $\sigma_{c,adm,\parallel}$ would be the permissible grade compression stress, parallel to the grain. (Note that the subscripts are separated by a comma.)

Modification factors

Table 3.1 gives the grade stresses and modulus of elasticity for the strength classes SC1 to SC9. The values given should be considered as basic ones which, when applied to a particular case, are factored to account for the difference between the basic case and the actual case in hand. The code lists over 90 *K* factors. However, for the purposes of this chapter the following *K* factors need only be considered:

K_1 (Table 3.4) for geometric properties in wet exposure conditions.

Table 3.3 Moisture content of timber for categories of end use

Position of timber in building	*Average moisture content attained in service conditions (%)*	*Moisture content which should not be exceeded at time of erection (%)*
External uses fully exposed	18 or more	–
Covered and generally unheated	18	24
Covered and generally heated	16	21
Internal in continuously heated building	14	19

K_2 (Table 3.5) for stresses and moduli in wet exposure conditions.
K_3 (Table 3.6) for duration of loading.
K_4 (Table 3.7 and Figure 3.1) for bearing stresses.
K_5 (Figure 3.2) for notched ends.
K_7 (Figure 3.3) for bending parallel to the grain.
K_8 for load sharing (see 'Load sharing' below).
K_{12} (Table 3.8) for members subject to axial compression.

Effective span

The effective span of a flexural member should be taken as the distance between the centres of bearing.

Table 3.4 Modification factor K_1 by which the geometrical properties of timber for the dry exposure condition should be multiplied to obtain values for the wet exposure condition

Geometrical property	*Value of K_1*
Thickness, width, radius of gyration	1·02
Cross-sectional area	1·04
First moment of area, section modulus	1·06
Second moment of area	1·08

Table 3.5 Modification factor K_2 by which dry stresses and moduli should be multiplied to obtain wet stresses and moduli applicable to wet exposure conditions

Property	*Value of K_2*
Bending parallel to grain	0·8
Tension parallel to grain	0·8
Compression parallel to grain	0·6
Compression perpendicular to grain	0·6
Shear parallel to grain	0·9
Mean and minimum modulus of elasticity	0·8

Table 3.6 Modification factor K_3 for duration of loading

Duration of loading	*Value of K_3*
Long term (e.g. dead + permanent imposed[a])	1·00
Medium term (e.g. dead + snow, dead + temporary imposed)	1·25
Short term (e.g. dead + imposed + wind[b], dead + imposed + snow + wind[b])	1·50
Very short term (e.g. dead + imposed + wind[c])	1·75

[a] For imposed floor loads K_3 = 1·00.
[b] For wind, short-term category applies to class C (15 s gust) as defined in CP3: Chapter V: Part 2.
[c] For wind, very short-term category applies to classes A and B (3 s or 5 s gust) as defined in CP3: Chapter V: Part 2.

Table 3.7 Modification factor K_4 for bearing stress

Length of bearing[a] (in mm)	10	15	25	40	50	75	100	150 or more
Value of K_4	1·74	167	1·53	1·33	1·20	1·14	1·10	1·00

[a] Interpolation is permitted.

Table 3.8 Modification factor K_{12} for compression members

$E/\phi_{c,\parallel}$	Value of K_{12}																			
Values of slenderness ratio λ (= L_e/i)	< 5	5	10	20	30	40	50	60	70	80	90	100	120	140	160	180	200	220	240	250
Equivalent L_e/b (for rectangular sections)	< 1·4	1·4	2·9	5·8	8·7	11·6	14·5	17·3	20·2	23·1	26·0	28·9	34·7	40·5	46·2	52·0	57·8	63·6	69·4	72·3
400	1·000	0·975	0·951	0·896	0·827	0·735	0·621	0·506	0·408	0·330	0·271	0·225	0·162	0·121	0·094	0·075	0·061	0·051	0·043	0·040
500	1·000	0·975	0·951	0·899	0·837	0·759	0·664	0·562	0·466	0·385	0·320	0·269	0·195	0·148	0·115	0·092	0·076	0·063	0·053	0·049
600	1·000	0·975	0·951	0·901	0·843	0·774	0·692	0·601	0·511	0·430	0·363	0·307	0·226	0·172	0·135	0·109	0·089	0·074	0·063	0·058
700	1·000	0·975	0·951	0·902	0·848	0·784	0·711	0·629	0·545	0·467	0·399	0·341	0·254	0·195	0·154	0·124	0·102	0·085	0·072	0·067
800	1·000	0·975	0·952	0·903	0·851	0·792	0·724	0·649	0·572	0·497	0·430	0·371	0·280	0·217	0·172	0·139	0·115	0·096	0·082	0·076
900	1·000	0·976	0·952	0·904	0·853	0·797	0·734	0·665	0·593	0·522	0·456	0·397	0·304	0·237	0·188	0·153	0·127	0·106	0·091	0·084
1000	1·000	0·976	0·952	0·904	0·855	0·801	0·742	0·677	0·609	0·542	0·478	0·420	0·325	0·255	0·204	0·167	0·138	0·116	0·099	0·092
1100	1·000	0·976	0·952	0·905	0·856	0·804	0·748	0·687	0·623	0·559	0·497	0·440	0·344	0·272	0·219	0·179	0·149	0·126	0·107	0·100
1200	1·000	0·976	0·952	0·905	0·857	0·807	0·753	0·695	0·634	0·573	0·513	0·457	0·362	0·288	0·233	0·192	0·160	0·135	0·116	0·107
1300	1·000	0·976	0·952	0·905	0·858	0·809	0·757	0·701	0·643	0·584	0·527	0·472	0·378	0·303	0·247	0·203	0·170	0·144	0·123	0·115
1400	1·000	0·976	0·952	0·906	0·859	0·811	0·760	0·707	0·651	0·595	0·539	0·486	0·392	0·317	0·259	0·214	0·180	0·153	0·131	0·122
1500	1·000	0·976	0·952	0·906	0·860	0·813	0·763	0·712	0·658	0·603	0·550	0·498	0·405	0·330	0·271	0·225	0·189	0·161	0·138	0·129
1600	1·000	0·976	0·952	0·906	0·861	0·814	0·766	0·716	0·664	0·611	0·559	0·508	0·417	0·342	0·282	0·235	0·198	0·169	0·145	0·135
1700	1·000	0·976	0·952	0·906	0·861	0·815	0·768	0·719	0·669	0·618	0·567	0·518	0·428	0·353	0·292	0·245	0·207	0·177	0·152	0·142
1800	1·000	0·976	0·952	0·906	0·862	0·816	0·770	0·722	0·673	0·624	0·574	0·526	0·438	0·363	0·302	0·254	0·215	0·184	0·159	0·148
1900	1·000	0·976	0·952	0·907	0·862	0·817	0·772	0·725	0·677	0·629	0·581	0·534	0·447	0·373	0·312	0·262	0·223	0·191	0·165	0·154
2000	1·000	0·976	0·952	0·907	0·863	0·818	0·773	0·728	0·681	0·634	0·587	0·541	0·455	0·382	0·320	0·271	0·230	0·198	0·172	0·160

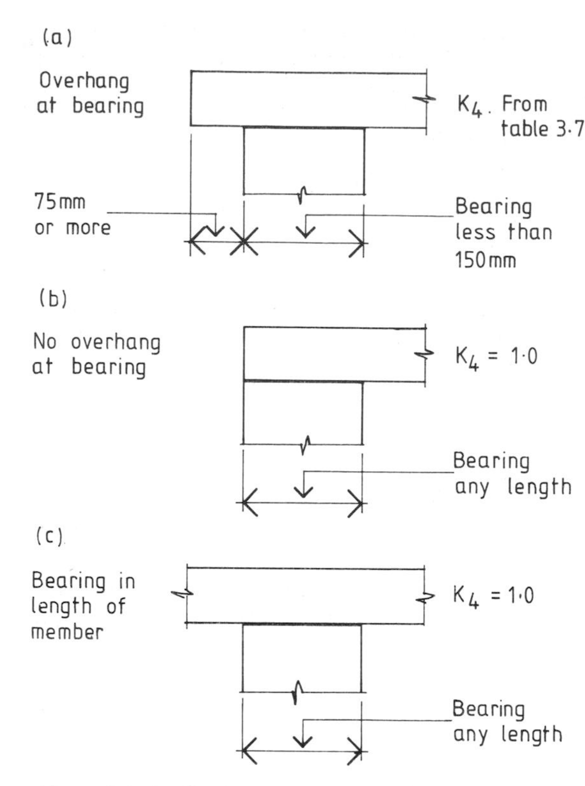

Figure 3.1 Application of modification factor K_4

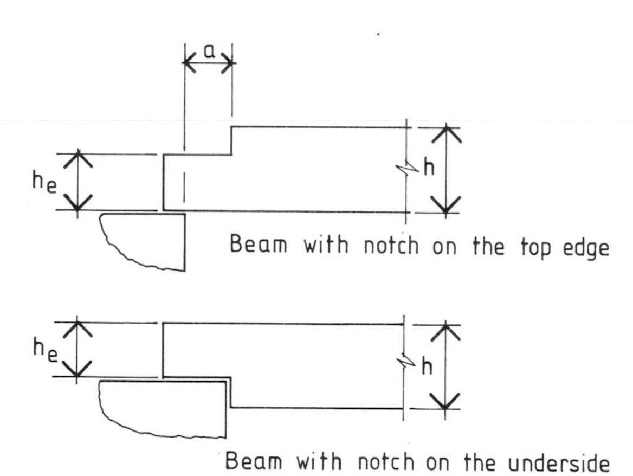

Figure 3.2 Modification factor K_S. The value of the modification factor, K_S, depends on the relative values of a and h_e for case (a) above and the ratio of h_e and h for case (b) above, as follows:

(a) $K_S = \dfrac{h(h_e - a) + ah_e}{h_e}$ for $a \leq h_e$

$K_S = 1{\cdot}0$ for $a > h_e$

(b) $K_S = \dfrac{h_e}{h}$

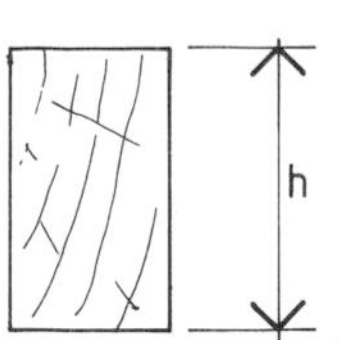

Figure 3.3 Modification factor K_7. The modification factor, K_7 for depth is to be taken as follows:

(a) $K_7 = 1{\cdot}17$ for $h \leq 72$ mm

(b) $K_7 = \left(\dfrac{300}{h}\right)^{0{\cdot}11}$ for $h > 72$ mm but < 300 mm

(c) $K_7 = 1{\cdot}0$ for $h \geq 300$ mm

Load sharing

Where four or more members such as joists in floors or compression members in stud walls, which share the applied loading and are not spaced further apart than 610 mm, with the load distributed by boarding, purlins, etc, the grade stresses in Table 3.1 should be multiplied by the load-sharing modification factor $K_8 = 1{\cdot}1$.

Deflection

For most general purposes, including domestic flooring, the deflection of the member when fully loaded should not exceed 0·003 of the span. For longer-span ($L > 4{\cdot}5$ m), domestic floors it should not exceed 14 mm. When calculating the actual deflection the following modulus of elasticity should be used:

1. Solid timber member acting alone; E_{min}
2. Load-sharing members; E_{mean}
3. Load sharing but subject to vibration; E_{min}.

Table 3.9 Maximum depth-to-breadth ratios (solid and laminated members)

Degree of lateral support	*Maximum depth-to-breadth ratio*
No lateral support	2
Ends held in position	3
Ends held in position and member held in line, as by purlins and tie rods	4
Ends held in position and compression edge held in line, as by direct connection of sheathing, deck or joists	5
Ends held in position and compression edge held in line, as by direct connection of sheathing, deck or joists, together with adequate bridging or blocking spaced at intervals not exceeding six times the depth	6
Ends held in position and both edges held firmly in line	7

Lateral support

Table 3.9 gives the maximum depth-to-breadth ratios for rectangular sections in bending. Members with ratios greater than those listed must be checked for buckling under design load.

3.2 A solid rectangular timber section in bending

1. Obtain the imposed and dead loading for the member (see Section 1.3, Chapter 1).
2. Calculate the maximum bending moment M and shearing force F_V values.
3. Establish whether it is part of a load-sharing system. If so, note that modification factor K_8 is to be used (i.e. $K_8 = 1{\cdot}1$).
4. Establish the service exposure condition. If wet, note modification factors K_1 and K_2 (Tables 3.4 and 3.5).
5. Establish duration of loading and note appropriate modification factor K_3 (Table 3.6).
6. Select a softwood species from Table 3.2 and note the appropriate strength class (SC) to obtain grade stresses from Table 3.1.
7. *Bending*. From Table 3.1 note the following grade stresses:
   (a) Bending parallel to grain: $\sigma_{m,\parallel}$
   (b) Shear parallel to grain: $\sigma_{\parallel}$
8. Calculate the section modulus Z_X required from:
   $M = Z_X \sigma_{m,adm,\parallel}$

where $\sigma_{m,adm,\parallel} = \sigma_{m,\parallel} \times$ appropriate modification factors from above excluding K_1 in 4, and

$$Z_{X_{reqd}} = \frac{M}{\sigma_{m,adm,\parallel}}$$

From Table 3.10 choose a section such that

$$Z_X > Z_{X_{reqd}}$$

(a starting point for floor joists would be $b = 47$ mm).

Remember: if there is a wet exposure condition then Z_X should be factored by K_1 (Table 3.4).

Table 3.10 Geometrical properties of planed all round softwoods: constructional timber

Finished size	Area	Section modulus, Z		Second moment of area, I		Radius of gyration, i	
		About XX	About YY	About XX	About YY	About XX	About YY
(mm)	(10^3 mm^2)	(10^3 mm^3)	10^3 mm^3)	(10^6 mm^4)	(10^6 mm^4)	(mm)	(mm)
33 × 72	2·38	28·5	13·1	1·03	0·216	20·8	9·53
33 × 97	3·20	51·7	17·6	2·51	0·260	28·0	9·53
33 × 120	3·96	79·2	21·8	4·75	0·359	34·6	9·53
33 × 145	4·79	116	26·3	8·38	0·434	41·9	9·53
35 × 72	2·52	30·2	14·7	1·09	0·257	20·8	10·1
35 × 97	3·40	54·9	19·8	2·66	0·347	28·0	10·1
35 × 120	4·20	84·0	24·5	5·04	0·429	34·6	10·1
35 × 145	5·08	123	29·6	8·89	0·518	41·9	10·1
35 × 169	5·92	167	34·5	14·1	0·604	48·8	10·1
35 × 194	6·79	220	39·6	21·3	0·693	56·0	10·1
35 × 219	7·67	280	44·7	30·6	0·782	63·2	10·1
41 × 72	2·95	35·4	20·2	1·28	0·414	20·8	11·8
41 × 97	3·98	64·3	27·2	3·12	0·557	28·0	11·8
41 × 120	4·92	98·4	33·6	5·90	0·689	34·6	11·8
41 × 145	5·95	144	40·6	10·4	0·833	41·9	11·8
41 × 169	6·93	195	47·3	16·5	0·971	48·8	11·8
41 × 194	7·95	257	54·4	24·9	1·11	56·0	11·8
41 × 219	8·98	328	61·4	35·9	1·26	63·2	11·8
41 × 244	10·0	407	68·4	49·6	1·40	70·4	11·8
41 × 294	12·1	591	82·4	86·8	1·69	84·9	11·8
44 × 72	3·17	38·0	23·2	1·37	0·511	20·8	12·7
44 × 97	4·27	69·0	31·3	3·35	0·689	28·0	12·7
44 × 120	5·28	106	38·7	6·34	0·852	34·6	12·7
44 × 145	6·38	154	46·8	11·2	1·03	41·9	12·7
44 × 169	7·44	209	54·5	17·7	1·20	48·8	12·7
44 × 194	8·54	276	62·6	26·8	1·38	56·0	12·7
44 × 219	9·64	352	70·7	38·5	1·55	63·2	12·7
44 × 244	10·7	437	78·7	53·3	1·73	70·4	12·7
44 × 294	12·9	634	94·9	93·2	2·09	84·9	12·7

Table 3.10 (*cont.*)

Finished size	*Area*	*Section modulus, Z*		*Second moment of area, I*		*Radius of gyration, i*	
(mm)	*(10^3 mm^2)*	*About XX (10^3 mm^3)*	*About YY 10^3 mm^3)*	*About XX (10^6 mm^4)*	*About YY (10^6 mm^4)*	*About XX (mm)*	*About YY (mm)*
47 × 72	3·38	40·6	26·5	1·46	0·623	20·8	13·6
47 × 97	4·56	73·7	35·7	3·57	0·839	28·0	13·6
47 × 120	5·64	113	44·2	6·77	1·04	34·6	13·6
47 × 145	6·82	165	53·4	11·9	1·25	41·9	13·6
47 × 169	7·94	224	62·2	18·9	1·46	48·8	13·6
47 × 194	9·12	295	71·4	28·6	1·68	56·0	13·6
47 × 219	10·3	376	80·6	41·1	1·89	63·2	13·6
47 × 244	11·5	466	89·8	56·9	2·11	70·4	13·6
47 × 294	13·8	677	108	99·5	2·54	84·9	13·6
60 × 97	5·82	94·1	58·2	4·56	1·75	28·0	17·3
60 × 120	7·20	144	72·0	8·64	2·16	34·6	17·3
60 × 145	8·70	210	87·0	15·2	2·61	41·9	17·3
60 × 169	10·1	286	101	24·1	3·04	48·8	17·3
60 × 194	11·6	376	116	36·5	3·49	56·0	17·3
60 × 219	13·1	480	131	52·5	3·94	63·2	17·3
72 × 97	6·98	113	83·8	5·48	3·02	28·0	20·8
72 × 120	8·64	173	104	10·4	3·73	34·6	20·8
72 × 145	10·4	252	125	18·3	4·51	41·9	20·8
72 × 169	12·2	343	146	29·0	5·26	48·8	20·8
72 × 194	14·0	452	168	43·8	6·03	56·0	20·8
72 × 219	15·8	576	189	63·0	6·81	63·2	20·8
72 × 244	17·6	714	211	87·2	7·59	70·4	20·8
72 × 294	21·2	1 040	254	152	9·14	84·9	20·8
97 × 97	9·41	152	152	7·38	7·38	28·0	28·0
97 × 145	14·1	340	227	24·6	11·0	41·9	28·0
97 × 194	18·8	608	304	59·0	14·8	56·0	28·0
97 × 244	23·7	962	383	117	18·6	70·4	28·0
97 × 294	28·5	1 400	461	205	22·4	84·9	28·0
145 × 145	21·0	508	508	36·8	36·8	41·9	41·9
145 × 194	28·1	910	680	88·2	49·3	56·0	41·9
145 × 294	42·6	2 090	1 030	307	74·7	84·9	41·9
194 × 194	37·6	1 220	1 220	118	118	56·0	56·0
244 × 244	59·5	2 420	2 420	295	295	70·4	70·4
294 × 294	86·4	4 240	4 240	623	623	84·9	84·9

[a] Measured at 20% moisture content. A tolerance of ± 0·5 mm is permitted.

9. Using Figure 3.3, calculate the appropriate K_7 modification factor value for the depth of the section n and adjust $\sigma_{m,adm,\parallel}$ value.
10. Recalculate $Z_{X_{reqd}}$ for the new permissible bending stress and check that

$$Z_X \geqslant Z_{X_{reqd}}$$

If so, the section is satisfactory for bending.

11. *Deflection*. Calculate the actual deflection δ_{act}. The actual deflection will be the sum of the deflection due to bending plus the deflection due to shear, i.e.

$$\delta_{act} = \delta_m + \delta_v$$

δ_m: see Table 1.9, Chapter 1, for typical load cases and δ_{max}.

$$\delta_v = \frac{\delta M}{AG}$$

where S = a shape factor, for a rectangular section ($S = 1{\cdot}2$), M = the maximum bending moment, A = the cross-sectional area ($= bh$) and G = the shear modulus = $0{\cdot}0625E$, where E = the appropriate elastic modulus.

For a rectangular section δ_v may be rewritten as

$$\delta_v = \frac{19{\cdot}2M}{AE}$$

Check:

(a) Short span, $L < 5{\cdot}6$ m:

$$\delta_{act} = \delta_m + \delta_v \leqslant 0{\cdot}003L$$

(b) Long span, $L \geq 5{\cdot}6$ m:

$$\delta_{act} = \delta_m + \delta_v \leq 0{\cdot}003L \leq 14 \text{ mm}$$

12. *Shear.* Calculate the actual shear stress v_{act}:

$$v_{act} = \left(\frac{F_V}{bh}\right) 1{\cdot}5$$

Note: see Figure 3.2 if the position being checked is notched.

13. Calculate the permissible shear stress parallel to the grain:

$$v_{adm,\parallel} = r_{\parallel} \times \text{modification factors } K_2, K_3, K_5 \text{ and } K_8 \text{ where appropriate}$$

and check that $v_{act} \leq v_{adm,\parallel}$.

13. *Bearing.* Establish the length and position of the bearing relative to the end of the member (refer to Figure 3.1 for guidance). From Table 3.1 read off appropriate grade stress for compression perpendicular to the grain, $\sigma_{c,\perp}$ and calculate the permissible stress

$$\sigma_{c,adm,\perp} = \sigma_{c,\perp} \times \text{modification factors, } K_2, K_3, K_4 \text{ and } K_8 \text{ where appropriate}$$

14. Calculate the actual bearing stress

$$\sigma_{act} = \frac{F_v}{bL_B}$$

where b = breadth of section and L_B = length of bearing, and check that $\sigma_{act} \leq \sigma_{c,adm,\perp}$.

3.3 A solid rectangular timber section in axial compression

1. Follow the steps 1–6 (excluding 2) as for a section in bending.
2. Establish the effective length L_E (see Table 3.11).
3. Calculate the slenderness ratio λ from either

$$\lambda = \frac{L_e}{b} \leq 52$$

where b = the least lateral dimension.

4. From Table 3.1 read off the appropriate grade stress for compression parallel to the grain $\sigma_{c,\parallel}$ and calculate the permissible stress:

$$\sigma_{c,adm,\parallel} = \sigma_{c,\parallel} \times \text{modification factors } K_2, K_3, K_8 \text{ and } K_{12} \text{ as appropriate}$$

Note: Modification factor K_{12} is used if $\lambda \geq 1{\cdot}4$.

From Table 3.8 calculate:

$$\frac{E_{min}}{\sigma_{c,\parallel} \times \text{modification factors } K_2, K_3 \text{ and } K_8 \text{ as appropriate}}$$

and using λ as calculated in 3, read off value of K_{12}.

5. Calculate the actual compressive stress:

$$\sigma_{c,act} = \frac{F}{A}$$

where F = the applied axial load and A = the sectional area, and check that

$$\sigma_{c,act} \leq \sigma_{c,adm,\parallel}$$

Table 3.11 Effective length of compression members

End conditions	*Effective length / Actual length* L_e/L
Restrained at both ends in position and in direction	0·7
Restrained at both ends in position and one end in direction	0·85
Restrained at both ends in position but not in direction	1·0
Restrained at one end in position and in direction and at the other end in direction but not in position	1·5
Restrained at one end in position and in direction and free at the other end	2·0

4

Reinforced concrete

4.1 Design requirements from BS 8110, *The structural use of concrete*

Many of the requirements of BS 8110 are too complicated to be summarized in a book of this description. Apart from the design of the more simple elements, reference should be made to the code itself.

Symbols

A_c area of concrete
A'_s area of compression reinforcement
A_{s_1} area of compression in the more highly compressed face
A_s area of tension reinforcement
A_{s_2} area of reinforcement in other face
A_{sc} area of longitudinal reinforcement (for columns)
A_{sv} cross-sectional area of the two legs of a link
a_b distance between bars
b_c breadth of compression face midway between restraints
b_t breadth of section at level of tension reinforcement
b_v width of section
b_w breadth of web or rib of a member
d effective depth of tension reinforcement
d' depth to compression reinforcement
d_2 depth to reinforcement
E_s modulus of elasticity of steel
e eccentricity
F ultimate load
F_k characteristic load
f_b bond stress
f_{cu} characteristic concrete cube strength
f_k characteristic strength
f_s service stress
f_y characteristic strength of reinforcement
f_{yv} characteristic strength of link reinforcement
G_k characteristic dead load
g_k characteristic dead load per unit area
h overall depth of section in plane of bending
h_f thickness of flange
k a constant (with appropriate subscripts)
l distance from face of support at end of a cantilever or effective span of a simply supported beam or slab
l_e effective height of a column or wall
l_{ex} effective height for bending about the major axis

l_{ey} effective height for bending about the minor axis
l_0 clear height of column between and restraints
l_x length of shorter side of a rectangular slab
l_y length of longer side of a rectangular slab
M bending moment due to ultimate loads
M_i maximum initial moment in a column due to ultimate loads
M_u ultimate resistance moment
N ultimate axial load at section considered
N_{bal} axial load on a column corresponding to the balanced condition
N_{uz} axial load capacity of a column ignoring all bending
n total ultimate load per unit area $(1{\cdot}4g_k + 1{\cdot}6q_k)$
Q_k characteristic imposed load
q_k characteristic imposed load per unit area
r internal radius of bend
s_b spacing of bars
s_v spacing of links along the member
V shear force due to ultimate loads
v shear stress
v_c ultimate shear stress in concrete
x neutral axis depth
z lever arm
γ_f partial safety factor for load
γ_m partial safety factor for strength
ΣA_{sv} area of shear reinforcement
Σu_s sum of the effective perimeters of the tension reinforcement
Φ bar size

Not all symbols associated with structural concrete are listed above. For a complete list reference should be made to BS 8110.

Limit state

BS 8110 defines limit state design as the achievement of acceptable probabilities that the structure being designed will not become unfit for the use for which it is required, i.e. that it will not reach a limit state. No attempt has been made in this book to derive the standard formulae used. The explanation of concepts inherent in limit state design can be found in many publications dealing with reinforced concrete design.

Characteristic loads

Since it is not yet possible to express loads in statistical terms, the following characteristic loads should be used in design:

1. *Dead loads.* The weight of the structure complete with finishes, partitions, etc.
2. *Imposed loads.* The weight due to furniture, occupants, etc.
3. *Wind loads.*

The following publications can be used to compute the loads on the structure:

BS 648, *Schedule of Weights of Building Materials.*
BS 6399, *Dead Loads and Imposed Loads.*

CP3: Chapter V: Part 2, *Wind loads*

Characteristic strength

This term is defined by BS 5328 as that value of the cube strength of concrete or the yield or proof stress of reinforcement below which not more than 5% of the test results fail. The characteristic strength is usually represented by the 28-day cube strength of the concrete and the yield of 0·2% proof stress of the reinforcement. Typical values of characteristic strength are given in Tables 4.1 and 4.2.

Table 4.1 Characteristic strength of reinforcement

Designation	*Nominal sizes (mm)*	f_y *(N/mm³)*
Hot-rolled mild steel	All sizes	250
High-yield steel	All sizes	460

The design may be based on f_y or a lower value if necessary to reduce deflection or control cracking.

Table 4.2 Characteristic strength of concrete

Concrete grade	*Characteristic compressive strength at 28 days (N/m²)*
C2·5	2·5
C5	5·0
C7·5	7·5
C10	10·0
C12·5	12·5
C15	15·0
C20	20·0
C25	25·0
C30	30·0
C35	35·0
C40	40·0
C45	45·0

Partial safety factors

1. *Loads.* In this case partial safety factors are introduced to take account of unforeseen variations in the characteristic loads. The partial safety factors differ for dead, imposed and wind loads and are set out in Table 4.3.

Table 4.3 Ultimate limit state

Load combination	Load type Dead		Imposed		Earth and water pressure	Wind
	Adverse	Beneficial	Adverse	Beneficial		
(1) Dead and imposed (and earth and water pressure)	1·4	1·0	1·6	0	1·4	–
(2) Dead and wind (and earth and water pressure)	1·4	1·0	–	–	1·4	1·4
(3) Dead, wind and imposed (and earth and water pressure)	1·2	1·2	1·2	1·2	1·2	1·2

The minimum applies when a reduction in load provides a more adverse condition (e.g. alternate spans in continuous systems for cases (1) and (2).

2. *Materials.* Such partial safety factors are introduced to allow for possible differences between the characteristic strength and the actual structural strength. The appropriate values of these factors are generally taken from Table 4.4.

Table 4.4 Values of γ_m for the ultimate limit state

Material	Factor
Reinforcement	1·15
Concrete in flexure or axial load	1·50
Shear strength without shear reinforcement	1·25
Bond strength	1·40
Others	>1·50

Design loads

The design load can be defined as the characteristic load times the partial safety factor. As previously mentioned, the partial safety factors vary according to the circumstances under which the loads are considered.

Effective span of beam

The effective span of a simply supported member should be taken as the smaller of:

1. The distance between the centres of bearings, or
2. The clear distance between supports plus the effective depth.

Slender beams

The clear distance between lateral restraints for a simply supported beam should not exceed $60b_c$ or $250b_c^2/d$, whichever is the lesser, for a cantilever beam with lateral restraint provided only at the support. The clear distance from the end of the cantilever to the face of the support should not exceed $25b_c$ or $100b_c^2/d$, whichever is the lesser.

Shear resistance of beams

1. Shear stress $v = V/b_v d$.
2. If v is greater than v_c the whole of the shearing force should be provided by shear reinforcement. In all cases v should not exceed $0{\cdot}8\sqrt{fc_u}$ or 5 N/mm^2, whichever is the lesser. The shear stress (v_c) which the concrete on its own can be allowed to resist is given in Table 4.5 for various percentages of bending reinforcement and various effective depths for 30 N/mm^2 concrete.
3. If v is less than v_c, nominal shear reinforcement should be provided throughout the span of the beam as noted in Table 4.6.
4. Where v is greater than v_c, shear reinforcement should be provided throughout the span of the beam as noted in Table 4.6. Tables 4.7 and 4.8 tabulate values that, when multiplied by the depth of the beam (mm), give the shear resistance for particular values of A_{sv} and s_v.
5. The spacing of vertical links in the direction of span (s_v) and at right angles to the span should not exceed $0{\cdot}75d$.
6. It is not necessary to provide shear reinforcement in slabs, bases, pile caps and similar members if v does not exceed v_c.
7. Up to 50% of shear reinforcement may be in the form of included bars (see BS 8110: Part 1).

Deflection of rectangular beams

For all normal cases the deflection of a beam will not be excessive if the ratio of its span to its effective depth is not greater than the appropriate ratio obtained from Table 4.9. The use of Table 4.9 will restrict the deflection to approximately 1/250 of the span. The table may be used for the

Table 4.5 Ultimate shear stresses for beams of various effective depths

$\frac{100A_s}{b_v d}$	Effective depth (mm)						
	150	*175*	*200*	*225*	*250*	*300*	*≥ 400*
≤ 0·15	0·46	0·44	0·43	0·41	0·40	0·38	0·36
0·25	0·54	0·52	0·50	0·49	0·48	0·46	0·42
0·50	0·68	0·66	0·64	0·62	0·59	0·57	0·53
0·75	0·76	0·75	0·72	0·70	0·69	0·64	0·61
1·00	0·86	0·83	0·80	0·78	0·75	0·72	0·67
1·50	0·98	0·95	0·91	0·88	0·86	0·83	0·76
2·00	1·08	1·04	1·01	0·97	0·95	0·91	0·85
≥ 3·00	1·23	1·19	1·15	1·11	1·08	1·04	0·97

f_{cu} = 30 N/mm²
For f_{cu} = 25 N/mm² the tabulated values should be divided by 1·062.
For f_{cu} = 35 N/mm² the tabulated values should be multiplied by 1·053.
For f_{cu} = 40 N/mm² the tabulated values should be multiplied by 1·10.

Table 4.6 Minimum provision of links in beam

Value of y (N/mm²)	*Area of shear reinforcement*
Less than $0{\cdot}5v_c$	Grade 250 (mild steel) links equal to 0·18% of the horizontal section throughout the beam, except in members of minor structural importance such as lintels
$0{\cdot}5v_c < v < (v_c + 0{\cdot}4)$	Minimum links for whole length of beam $A_{sv} > \frac{0{\cdot}4b_w S_v)}{0{\cdot}87f_{yv}}$
$(v_c + 0{\cdot}4) < v$	Links only provided $A_{sv} > b_w \frac{S_v(v - v_c)}{0{\cdot}87f_{yv}}$

Table 4.7 Shear resistance of links with f_{yv} – 250 N/mm² and $v > (v_c + 0{\cdot}4)$. (Values in N/mm depth of beam)

s_v	Diameters				
	6	*8*	*10*	*12*	*16*
75	164	293	456	657	1168
100	123	220	342	492	876
125	98	176	273	394	702
150	82	146	228	328	584
175	70	125	195	281	501
200	61	110	171	246	438
250	49	88	136	197	351
300	41	73	114	164	292
350	35	63	98	141	250
400	31	55	85	123	219
450	26	49	76	109	195
500	25	44	68	98	175
550	22	40	62	89	159
600	21	37	57	82	146
700	18	31	49	70	125

Table 4.8 Shear resistance of links with f_{yv} – 460 N/mm² and $v > (v_c + 0{\cdot}4)$. (Values in N/mm depth of beam)

s_v	Diameters				
	6	*8*	*10*	*12*	*16*
75	302	539	839	1209	2149
100	226	405	629	905	1612
125	180	324	502	725	1292
150	151	269	420	604	1075
175	129	230	359	517	922
200	112	202	315	453	806
250	90	162	250	362	646
300	75	134	210	302	537
350	64	116	180	259	460
400	57	101	156	226	403
450	48	90	140	201	359
500	46	81	125	180	322
550	40	74	114	164	293
600	39	68	105	151	269
700	33	57	90	129	230

calculations relating to beams with a span of more than 10 m if the deflection is acceptable. Otherwise it is necessary to use the values from Table 4.9 multiplied by 10/span except for cantilevers, when the design should be justified by calculation. Values of span/effective depth ratio should be multiplied by the appropriate factor obtained from Tables 4.10 or 4.11.

Table 4.9 Basic span/effective depth ratios for rectangular or flanged beams

Support conditions	*Rectangular beams*	*Flanged beams with $b_w/b < 0{\cdot}3$*
Cantilever	7	5·6
Simply supported	20	16·0
Continuous	26	20·8

Table 4.10 Modification factor for tension reinforcement

Service stress		M/bd^2								
		0·50	0·75	1·00	1·50	2·00	3·00	4·00	5·00	6·00
	100	2·00	2·00	2·00	1·86	1·63	1·36	1·19	1·08	1·01
	150	2·00	2·00	1·98	1·69	1·49	1·25	1·11	1·01	0·94
(f_y = 250)	156	2·00	2·00	1·96	1·66	1·47	1·24	1·10	1·00	0·94
	200	2·00	1·95	1·76	1·51	1·35	1·14	1·02	0·94	0·88
	250	1·90	1·70	1·55	1·34	1·20	1·04	0·94	0·87	0·82
(f_y = 460)	288	1·68	1·50	1·38	1·21	1·09	0·95	0·87	0·82	0·78
	300	1·60	1·44	1·33	1·16	1·06	0·93	0·85	0·80	0·76

Note: The design service stress may be estimated from the equation obtained from BS 8110: Part 1.

Table 4.11 Modification factor for compression reinforcement

$\frac{100A'_s\text{ (prov)}}{b_d}$	Factor
0·00	1·00
0·15	1·05
0·25	1·08
0·35	1·10
0·50	1·14
0·75	1·20
1·00	1·25
1·50	1·33
2·00	1·40
2·50	1·45
> 3·00	1·50

Note: The area of compression reinforcement A'_s (prov) used in this table may include all bars in the compression zone, even those not effectively tied with links.

The minimum effective depth d of a rectangular beam with tension reinforcement only that will comply with the requirements of BS 8110 may be written as:

$$d = \frac{\text{Effective span}}{(\text{factor, Table 4.9}) \times (\text{factor, Table 4.10})}$$

The minimum effective depth d of a rectangular beam with tension and compression reinforcement that will comply with the requirements of BS 8110 may be written as:

$$d = \frac{\text{Effective span}}{(\text{factor, Table 4.9}) \times (\text{factor, Table 4.10}) \times (\text{factor, Table 4.11})}$$

Deflection of flanged beams

For a flanged beam the span/effective depth ratio may be determined as for rectangular beams. However, for values of b_w/b greater than 0·3, linear interpolation between the values given in Table 4.9 for rectangular beams and for flanged beams with b_w/b, 0·3 may be used.

Anchorage bond stress

Recommendations for the design of anchorage bond stresses are shown in BS 8110: Part 1.

Local bond stress

Providing that the force in a bar can be developed by the appropriate anchorage length, local bond stress may be ignored.

Laps in bars

The length should be at least equal to the design anchorage length necessary to develop the required stress in the tension reinforcement, but in the case of compression reinforcement it should be at least 25% greater than the design anchorage compression length. In both cases lap lengths for bars of unequal size (or wires in fabric) may be based upon the smaller bar.

The following points should also be noted.

1. Where a lap occurs at the top of a section as cast and the minimum cover is less than twice the size of the lapped reinforcement, the lap length should be increased by a factor of 1·4.
2. Where a lap occurs at the corner of a section and the minimum cover to either face is less than twice the size of the lapped reinforcement, or where the clear distance between adjacent laps is less than 75 mm or six times the size of the lapped reinforcement, whichever is the greater, the lap length should be increased by a factor of 1·4.
3. In cases where conditions 1 and 2 both apply the lap length should be increased by a factor of 2·0. Values for lap lengths are given in Table 4.12 as multiples of bar size.

Table 4.12 Ultimate anchorage bond and lap lengths

f_{cu} *(N/mm²)*	*25*			*30*			*40 or over*		
Reinforcement type	250	460[a]	Fabric[b]	250	460[a]	Fabric[b]	250	460[a]	Fabric[b]
Tension anchorage and lap lengths	39	41	31	36	37	29	31	32	25
1·4 × tension lap	55	57	44	50	52	40	43	45	35
2·0 × tension lap	78	81	62	71	74	57	62	64	49
Compression anchorage length	32	32	25	29	29	23	25	26	20
Compression lap length	39	40	31	36	37	29	31	32	25

[a] Deformed bars type 2.
[b] Welded fabric complying with BS 4483.

Hooks and bends

The effective anchorage length of a hook or bend should be measured from the start of the bend to a point four times the bar size beyond the end of the bend. This may be taken as the lesser of 24 times the bar size, or

1. For a hook – eight times the internal radius of the hook or the active length of the bar in the hook, including the straight portion whichever is greater.
2. For a bend – four times the internal radius of the bend with a maximum of 12 times the bar size, or the actual length of the bar whichever is greater.

Curtailment of bars

In any member subject to bending, every curtailed bar should extend (except at end supports) beyond the calculated cut-off point for a distance equal to the effective depth of the member or 12 times the bar size, whichever is greater. In addition, bars should not be stopped off in a tension zone, unless one of the following conditions is satisfied:

1. The bars extend an anchorage length appropriate to their design strength ($0{\cdot}87f_y$) from the point at which it is no longer required to assist in resisting the bending moment.
2. The shear capacity of the section, where the reinforcement stops, provide double the area required for the moment at such points.

Anchorage of bars

At a simply supported end of a member one of the following requirements should be fulfilled:

1. Effective anchorage equivalent to 12 bar sizes beyond the centre line of support. (No hook or bend should begin before the centre of the support.)
2. Effective anchorage equivalent to 12 bar sizes plus $d/2$ from the face of the support. (No bend should begin $d/2$ from the face of the support.)
3. For slabs, if the design ultimate shear stress at the face of the support is less than half the appropriate value v_c recommended in Table 4.5 a straight length of bar beyond the centre-line of the support equal to either one third of the support width, or 30 mm, whichever is greater.

Simplified rules for the curtailment of bars are given in BS 8110: Part 1.

Cover

General rules for reinforcement covers are given in BS 8110: Part 1. Table 4.13 gives the normal range of conditions of exposure and the required nominal cover to the reinforcement.

In the above context bar size can be defined as follows:

1. Individual bars – not less than the diameter of the bar.
2. Pairs or bundles of bars – not less than the diameter of a single bar of equivalent area.

Minimum distance between bars

The lateral dimension between bars should be the maximum-sized aggregate plus 5 mm or the bar size, whichever is greater. Vertical dimension between bars should be two-thirds of the maximum-sized aggregate.

Maximum distance between bars in tension

General rules for establishing the maximum distance between bars in tension are given in BS 8110: Part 1.

Table 4.13 Nominal cover to all reinforcement (including links)

Conditions of exposure	*Nominal cover (mm)*				
Mild	25	20	20[a]	20[a]	20[a]
Moderate	–	35	30	25	20
Severe	–	–	40	30	25
Very severe	–	–	50[b]	40[b]	30
Extreme	–	–	–	60[b]	50
Maximum free water/cement ratio	0·65	0·60	0·55	0·50	0·45
Minimum cement content (kg/m^3)	275	300	325	350	400
Lowest grade of concrete	C30	C35	C40	C45	C50

[a] These covers may be reduced to 15 mm provided that the nominal size of aggregate does not exceed 15 mm.
[b] Where concrete is subject to freezing while wet, air entrainment should be used.

Minimum and maximum percentages of reinforcement in members

The minimum and maximum percentages of reinforcement appropriate for various conditions of loading and types of member are given in BS 8110: Part 1.

Reinforcement area tables

Reinforced concrete design calculations usually end with an area of reinforcement required.

For columns, beams and other similar structural members, the reinforcement required at any section will be quoted in mm^2. This has to be translated into a certain number of bars of given diameter. If, for example, a column required 2500 mm^2 of reinforcement then it can be seen from Table 4.14 that eight 25 mm bars would be suitable, as their total cross-sectional area is 2510 mm^2.

In the case of slabs and other similar structural forms, the required reinforcement will be specified in mm^2 per metre width. For example, if a slab required 600 mm^2/m then, from Table 4.15, 12 mm bars at 175 mm centre to centre would seem appropriate, as their total cross-sectional area is 646 mm^2/m width.

4.2 Basic principles for the design of solid slabs and rectangular beams

Assumptions

1. Plane sections remain plane in bending.
2. The tensile strength of concrete is ignored.
3. The stress distribution in the concrete in compression is derived from the stress–strain curve shown in BS 8110: Part 1.
4. The stress in the reinforcement is derived from the stress–strain curve shown in BS 8110: Part 1.

Ultimate moments of resistance (based on tension reinforcement only)

M_u (concrete) $= K'f_{cu}bd^2$ (where $K' = 0{\cdot}156$)

M_u (tension reinforcement) $= (0{\cdot}87f_y)\,A_sz$

These moments of resistance have been prepared

Table 4.14 Areas of reinforcement for beams and columns

Dia. (mm)	*Mass (kg/m)*	*Areas in mm^2 for groups of bars*											
		1	*2*	*3*	*4*	*5*	*6*	*7*	*8*	*9*	*10*	*11*	*12*
6	0·222	28	57	85	113	142	170	198	226	255	283	311	340
8	0·395	50	101	151	201	252	302	352	402	453	503	553	604
10	0·616	79	157	236	314	392	471	550	628	707	785	864	942
12	0·888	113	226	339	452	566	679	792	905	1 020	1 130	1 240	1 360
16	1·579	201	402	603	804	1 010	1 210	1 410	1 610	1 810	2 010	2 210	2 410
20	2·466	314	628	943	1 260	1 570	1 890	2 200	2 510	2 830	3 140	3 460	3 770
25	3·854	491	982	1 470	1 960	2 450	2 950	3 440	3 930	4 420	4 910	5 400	5 890
32	6·313	804	1 610	2 410	3 220	4 020	4 830	5 630	6 430	7 240	8 040	8 850	9 650
40	9·864	1 260	2 510	3 770	5 030	6 280	7 540	8 800	10 100	11 300	12 600	13 800	15 100
5?	15·413	1 960	3 930	5 890	7 850	9 820	11 800	13 700	15 700	17 700	19 600	21 600	23 600

Table 4.15 Areas of reinforcement for slabs and walls

Dia. (mm)	*Areas in mm² for centres in mm*									*Perimeter (mm)*
	50	75	100	125	150	175	200	250	300	
6	566	377	283	226	189	162	142	113	94	18·9
8	1010	671	503	402	335	287	252	201	168	25·2
10	1570	1050	785	628	523	449	393	314	262	31·4
12	2260	1510	1130	905	745	646	566	452	377	37·6
16	4020	2680	2010	1610	1340	1150	1010	804	670	50·3
20	6280	4190	3140	2510	2090	1800	1570	1260	1050	62·8
25	9820	6550	4910	3930	3270	2810	2450	1960	1640	78·5
32	16100	10700	8040	6430	5360	4600	4020	3220	2680	100·0
40	25100	16800	12600	10100	8380	7180	6280	5030	4190	125·6
50	39200	26200	19600	15700	13100	11200	9800	7850	6550	157·0

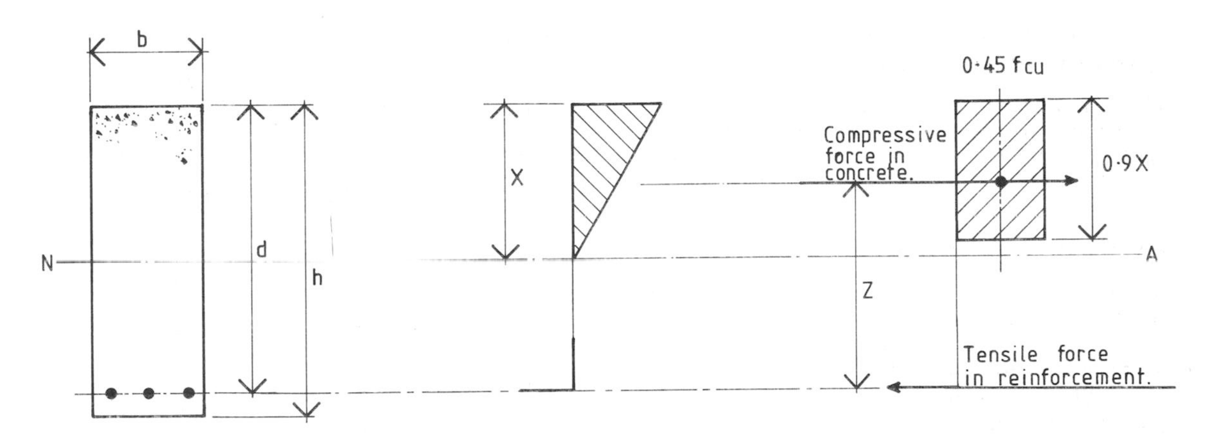

Figure 4.1 Beam properties

using the simplified stress block as shown in Figure 4.1. The maximum value of z is $0{\cdot}775d$ in this case. The moment of resistance of the section at the ultimate limit state must be equal to or greater than the ultimate applied bending moment.

Basis of design

To find the amount of tension reinforcement in a beam the following methods can be used:

1. Design charts.
2. Design formulae.

Design charts have been prepared using the stress–strain curves for concrete and steel as shown in BS 8110: Part 1. These charts, which are to be found in BS 8110: Part 3, are prepared for a particular grade of concrete and strength of reinforcement and a typical chart is shown in Figure 4.2.

If $M < m_u$ for the concrete, the area of tension

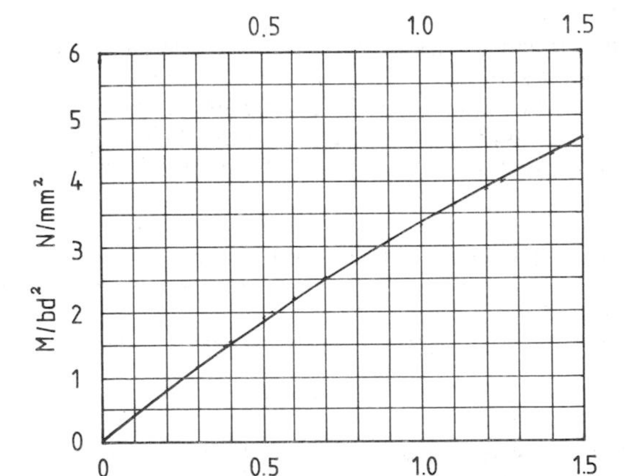

Figure 4.2 Typical design chart for beams

Table 4.16 Lever arm and neutral axis depth factors for beams

$K = \dfrac{M}{f_{cu}bd^2}$	0·05	0·06	0·07	0·08	0·09	0·100	0·110	0·120	0·130	0·140	0·150	0·156
$a = \dfrac{z}{d}$	0·94	0·93	0·91	0·90	0·89	0·87	0·86	0·84	0·82	0·80	0·79	0·775
$n = \dfrac{x}{d}$	0·13	0·16	0·19	0·22	0·25	0·29	0·32	0·35	0·39	0·43	0·47	0·50

reinforcement A_s can be calculated from:

$A_s = M/(0{\cdot}87 f_y)z$

where z is obtained from Table 4.16 for various values of K.

4.3 Simple slab design

Simply supported slab spanning in one direction

1. Decide on the material stresses to be used, i.e. f_{cu} and f_y.
2. Assume overall thickness h of slab.
3. Estimate the characteristic loads g_k and q_k per unit area.
4. Calculate the design loads $1{\cdot}4g_k$ and $1{\cdot}6q_k$ per unit area.
5. Determine the ultimate bending moment M.
6. Choose the appropriate concrete cover from Table 4.13 and determine d.
7. Check the cover is appropriate for fire resistance if necessary (see BS 8100: Part 1).
8. Determine the ultimate resistance moment based on the concrete section (this must be equal to or greater than the ultimate bending moment or a larger section must be considered).
9. Check the span/depth ratio:
   (a) Calculate M/bd^2 and obtain modification factor from Table 4.10;
   (b) Select the basic ratio from Table 4.9;
   (c) Calculate the allowable ratio from basic ratio × modification factor;
   (d) Actual ratio = effective span/effective depth;
   (e) Actual ratio < allowable ratio.
10. Calculate the reinforcement area by either:
    (a) Use of design charts: in which case, calculate M/bd^2 and determine $100A_s/bd$ from Figure 4.2 or appropriate design chart in BS 8100: Part 3; or
    (b) Use of design formula: in which case, calculate $A_s = M/(0{\cdot}87f_y)z$.
11. Select the reinforcement required from Table 4.15.
12. Check the minimum area of reinforcement in either direction, which should not be less than the following: $0{\cdot}0013bh$ for high-yield steel; $0{\cdot}0024bh$ for mild steel, where $b = 1000$ mm and h = overall depth of slab (mm).
13. The bar spacing (maximum) is to be in accordance with BS 8100: Part 1.

Simply supported slab spanning in two directions

1. Decide on the material stresses to be used, i.e. f_{cu} and f_y.
2. Assume overall thickness h of slab.
3. Estimate the characteristic loads g_k and q_k per unit area.
4. Calculate design load $n = 1{\cdot}4g_k + 1{\cdot}6q_k$.
5. Determine the ultimate bending moments in the short and long spans from:
   Short span: $M_{sx} = \alpha_{sx} n l_x^2$
   Long span: $M_{sy} = \alpha_{sy} n l_x^2$

where n = the total ultimate load per unit area ($1{\cdot}4g_k + 1{\cdot}6q_k$),
l_y = the length of the longer side,
l_x = the length of the shorter side,
α_{sx} and α_{sy} = the moment coefficients from Table 4.17,

Table 4.17 Bending moment coefficients for slabs spanning in two directions, at right angles, simply supported on four sides

$\dfrac{l_y}{l_x}$	1·0	1·1	1·2	1·3	1·4	1·5	1·75	2·0	2·5	3·0
α_{sx}	0·062	0·074	0·084	0·093	0·099	0·104	0·113	0·118	0·122	0·124
α_{sy}	0·062	0·061	0·059	0·055	0·051	0·046	0·037	0·029	0·020	0·014

At least 50% of the tension reinforcement provided at midspan should extend to the supports. The remaining 50% should extend to within 0·1 or 0·1 of the support, as appropriate.

M_{sx} and M_{sy} = the moments at midspan on strips of unit width and spans l_x and l_y.

6. Choose the appropriate concrete cover from Table 4.13 and determine d.
7. Check the cover is appropriate for fire resistance if necessary (see BS 8100: Part 1).
8. Determine the ultimate resistance moment based on the concrete section for the short span (this must be equal to or greater than the ultimate bending moment for the short span or a larger section must be considered).
9. Check the span/depth ratio:
   (a) Calculate M/bd^2 and obtain modification factor from Table 4.10;
   (b) Select the basic ratio from Table 4.9;
   (c) Calculate the allowable ratio from basic ratio × modification factor;
   (d) Actual ratio = effective short span/effective depth;
   (e) Actual ratio < allowable ratio.
10. Calculate the reinforcement areas by either:
    (a) Use of design charts: in which case, calculate M_{sx}/bd^2 and M_{sy}/bd^2, and determine $100A_s/bd$ from Figure 4.2 or the appropriate design chart in BS 8100: Part 3; or
    (b) Use of design formula: in which case calculate $A_s = M_{sx}/(0{\cdot}87f_y)z$ and $A_s = M_{sy}/(0{\cdot}87f_y)z$.

    (*Note:* Care should be taken to use the value of d appropriate to the direction of the reinforcement (see Figure 4.3).)
11. Select the reinforcement required from Table 4.15.
12. Check the minimum area of reinforcement in either direction should not be less than the following: $0{\cdot}0013bh$ for high-yield steel; $0{\cdot}0024bh$ for mild steel, where b = 1000 mm and h = overall depth of slab(mm).

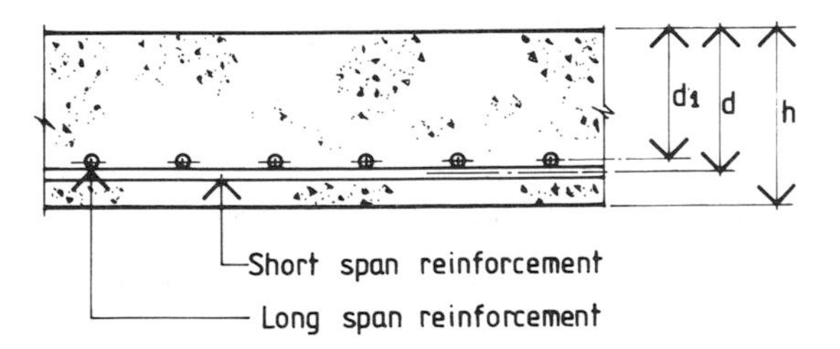

Figure 4.3 Two-way spanning slab

13. The bar spacing (maximum) is to be in accordance with BS 8100: Part 1.

4.4 Simple beam design

Simply supported rectangular beam with tension reinforcement only

1. Decide on the material stresses to be used, i.e. f_{cu} and f_y.
2. Assume beam size (d = effective span/12, $b = d/2$).
3. Estimate the characteristic loads g_k and q_k per unit length of beam.
4. Calculate the design loads $1{\cdot}4g_k$ and $1{\cdot}6q_k$ per unit length of beam.
5. Determine the ultimate bending moment M.
6. Choose the appropriate concrete cover from Table 4.13 and determine the overall depth of beam.
7. Check the cover is appropriate for fire resistance if necessary (see BS 8100: Part 1).
8. Determine the ultimate resistance moment based on the concrete section (this must be equal to or greater than the ultimate bending moment, otherwise a larger section or compression reinforcement must be considered).
9. Check the span/depth ratio:
   (a) Calculate m/bd^2 and obtain the modification factor from Table 4.10;
   (b) Select the basic ratio from Table 4.9;
   (c) Calculate the allowable ratio from basic ratio × modification factor.
   (d) Actual ratio = effective span/effective depth.
   (e) Actual ratio < allowable ratio.
10. Calculate the reinforcement area by either:
    (a) Use of design charts: in which case, calculate M/bd^2 and determine $100A_s/bd$ from Figure 4.2 or appropriate design chart in BS 8100: Part 3; or
    (b) Use of design formula: in which case, calculate $A_s = M/(0{\cdot}87f_y)z$.
11. Select the tension reinforcement required from Table 4.14.
12. Check the minimum and maximum reinforcement areas from BS 8100: Part 1.
13. Calculate the shear reinforcement:
    (a) Calculate the maximum shear force V;
    (b) Calculate the actual shear stress v from $V/b_v d$;
    (c) Determine $100A_s/bd$ (A_s = minimum tension reinforcement allowing for curtailments);
    (d) Select v_c from Table 4.5. If $v_c + 0{\cdot}4 < v$, provide shear reinforcement;
    (e) Determine the bar size and spacing for nominal links from Table 4.6;

(f) Shear resistance of beam = shear resistance of concrete + shear resistance of nominal links = $v_c b_v d$ + value from Table 4.7 or 4.8 × effective depth of beam;
(g) If the shear resistance of the beam is less than the ultimate applied shearing force V, draw a shear force diagram and work out stopping of points for shear links (see Figure 4.4);
(h) Calculate $b_v(v - v_c)$ and select from Tables 4.7 or 4.8 a suitable bar size and spacing of links.

14. Recommendations for the design of anchorage bond stress and local bond stress are given in BS 8100: Part 1.

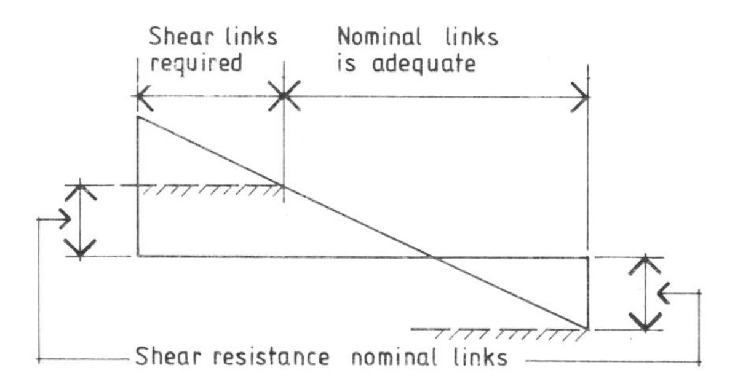

Figure 4.4 Shear force diagram

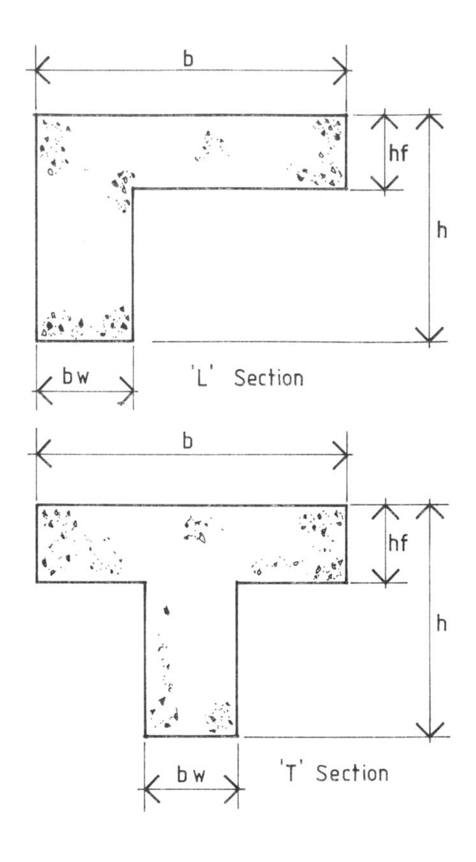

h = Overall depth
bw = Breadth of rib (or web)
b = Effective width
hf = Overall depth of slab (or flange)

Figure 4.5 Flanged beams

4.5 Basic principles for the design of simply supported flanged beams with tension reinforcement only

Introduction

Most beams form part of a floor arrangement in which the reinforced concrete floor is monolithic with the beam. The floor acts as the compression zone for the beam, thus increasing the load capacity of the member. Details of the two most common types of flanged beams are shown in Figure 4.5.

Effective width of flange

For a T-beam it is the lesser of:

1. The width of the rib + 0·2 × the distance between points of zero moment.
2. The actual width of the flange.

For an L-beam it is the lesser of:

1. The width of the rib + 0·1 × the distance between points of zero moment.
2. The actual width of the flange.

Assumptions

These are stated in Section 4.2 but are repeated here.

1. Plane sections remain plane in bending.
2. The tensile strength of the concrete is ignored.
3. The stress distribution in the concrete in compression is derived from the stress–strain curve shown in BS 8100: Part 1.
4. The stress in the reinforcement is derived from the stress–strain curve shown in BS 8100: Part 1.

Basis of design

To find the amount of reinforcement in a flanged beam one of the following methods can be used:

1. Design charts.
2. Design formulae.

Design charts may be used for the design of

flanged beams providing the depth to the neutral axis lies within the thickness of the flange. When the neutral axis lies outside the thickness of the flange the design formula method should be used.

4.6 Simple flanged beam design

Simply supported flanged beam with tension reinforcement only

1. Decide on the material stresses to be used, i.e. f_{cu} and f_y.
2. Assume an overall depth and breadth of web.
3. Estimate the characteristic loads g_k and q_k per unit length of beam.
4. Calculate the design loads $1{\cdot}4g_k$ and $1{\cdot}6q_k$ per unit length of the beam.
5. Determine the ultimate bending moment M.
6. Calculate the maximum permissible width of the flange.
7. Choose the appropriate concrete cover from Table 4.13 and determine d.
8. Check the position of the neutral axis by determining $K = M/f_{cu}bd^2$, where b = width of flange.
9. Select values of n and z from Table 4.16 and calculate x.
10. (a) If $0{\cdot}9x > h_f$ then the neutral axis lies within the flange and A_s can be determined as for a rectangular beam.
    (b) If $0{\cdot}9x < h_f$ then the neutral axis lies outside the flange: Calculate the ultimate resistance moment of the flange M_{uf} from $M_{uf} = 0{\cdot}45f_{cu}(b - b_w)h_f(d - 0{\cdot}5h_f)$, where b_w = breadth of the web.
    (c) Calculate $K_t = M - M_{uf}/f_{cu}b_w d^2$. If $K_t < K'$ obtained from Table 4.16 then select the value of a from the table. If $K_t > K'$, redesign the section or consult BS 8100 for the design of compression steel.
11. Check span/depth ratio:
    (a) Calculate M/bd^2 and obtain the modification factor from Table 4.10 (b is the effective width of the flange).
    (b) Select the basic ratio from Table 4.9.
    (c) Calculate the allowable ratio from basic ratio × modification factor.
    (d) Actual ratio = effective span/effective depth.
    (e) Actual ratio < allowable ratio.
12. Calculate the reinforcement area by either:

    $A_s = M/(0{\cdot}87f_y)z$; or

    $A_s = M_{uf}/0{\cdot}87f_y\ (d - 0{\cdot}5h_f) + M - M_{uf}/(0{\cdot}87f_y)z$

    depending on the position of the neutral axis.
13. Select the tension reinforcement required from Table 4.14.
14. Check the minimum and maximum reinforcement areas from BS 8100: Part 1.
15. Calculate shear and nominal links (see Section 4.4). The beam breadth in this case is taken as the breadth of web b_w.
16. Recommendations for the design of anchorage bond stress and local bond stress are given in BS 8100: Part 1.

4.7 Simply supported hollow block slab design

Introduction

BS 8100 defines this type of slab as a series of concrete ribs, cast *in situ*, between blocks which remain part of the complete structure (Figure 4.6). The tops of the ribs are usually connected by a concrete topping of the same strength as that used in the ribs. Although it is suggested that under certain stated conditions the blocks contribute to the thickness of the structural topping and breadth of the rib, in the treatment given here the hollow blocks are not regarded as contributing to the structural strength and slip tiles are not considered. As the size and weight of the blocks vary, the accurate weight of floor should be worked out from the trade literature.

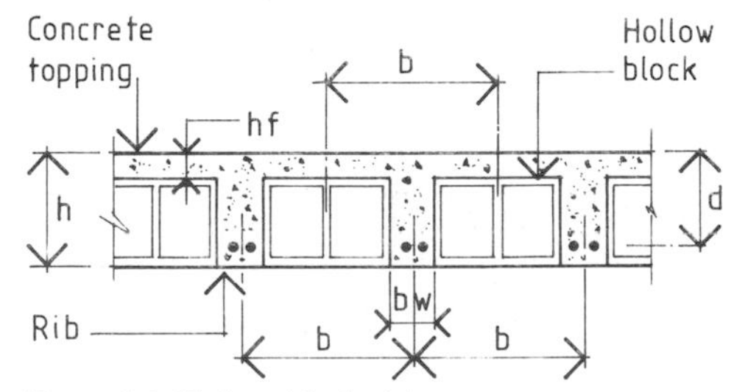

Figure 4.6 Hollow block slab

Design

1. Decide on the material stresses to be used, i.e. f_{cu} and f_y.
2. Assume a size of block, thickness of structural topping and breadth of rib.
3. Calculate the maximum permissible width of flange (usually taken as the centres of the ribs).
4. Calculate the characteristic loads g_k and q_k per unit length of slab per width of flange.
5. Calculate the design loads $1{\cdot}4g_k$ and $1{\cdot}6q_k$ per unit length of slab per width of flange.
6. Determine the ultimate bending moment M.
7. Choose the appropriate concrete cover from Table 4.13 and determine d.
8. Check the overall thickness is appropriate for fire resistance if necessary (see BS 8100: Part 1).

9. Check the position of the neutral axis (see Section 4.6).
10. Select values of n and z from Table 4.16 and calculate x.
11. Calculate the ultimate resistance moment of the section (see Section 4.6).
12. Check span/depth ratio (see Section 4.6).
13. Calculate the reinforcement area by either:

    $A_s = M/(0{\cdot}87f_y)z$; or $A_s = M_{uf}/(0{\cdot}87f_y)(d - 0{\cdot}5h_f) + M - M_{uf}/(0{\cdot}87f_y)z$, depending upon the position of the neutral axis.
14. Select the tension reinforcement required from Table 4.14.
15. Check the minimum and maximum reinforcement areas from BS 8100: Part 1.
16. Check the shear stress. The critical point for shear is usually at the edge of the hollow block and solid concrete section. The shear will be taken on the rib only. BS 8100 recommends that shear reinforcement is not provided in slabs.
    (a) Calculate the shear force V at the edge of the hollow block section.
    (b) Determine the actual shear stress v.
    (c) Determine the allowable shear stress v_c.
    (d) If $v_c > v$, then the slab is suitable.
    (e) If $v_c < v$, it may be necessary to adjust the critical point for shear so that the value of v is decreased below that of v_c.
17. Recommendations for the design of anchorage bond stress and local bond stress are given in BS 8100: Part 1.

4.8 Simple column base design

Introduction

The thickness of the base must be sufficient to resist the shearing forces and bending moments safely. The allowable bearing pressure of the soil under the base is normally determined from tests on soil samples. However, BS 8004 provides a guide to the allowable pressures which can be used for the preliminary design of the base. The size of the base should be determined using the column 'serviceability' loading. The design of the reinforcement for the base should be carried out using ultimate limit state.

Design of a square base with concentric column load only (Figure 4.7)

Symbols used in the design

A = plan area of base ($L \times L$)
A_1 = area within the critical shear perimeter
d = effective depth of tension reinforcement
h = overall thickness of base
N_s = axial column load (service)
N_d = axial column load (design)
N_1 = load under transverse shear
N_2 = load within area of critical perimeter
N_3 = load against punching
M = ultimate bending moment
p = critical shear perimeter
q_s = allowable bearing pressure of soil (service)
q_u = actual bearing pressure (ultimate)
v_1 = actual transverse shear
v_2 = actual punching shear

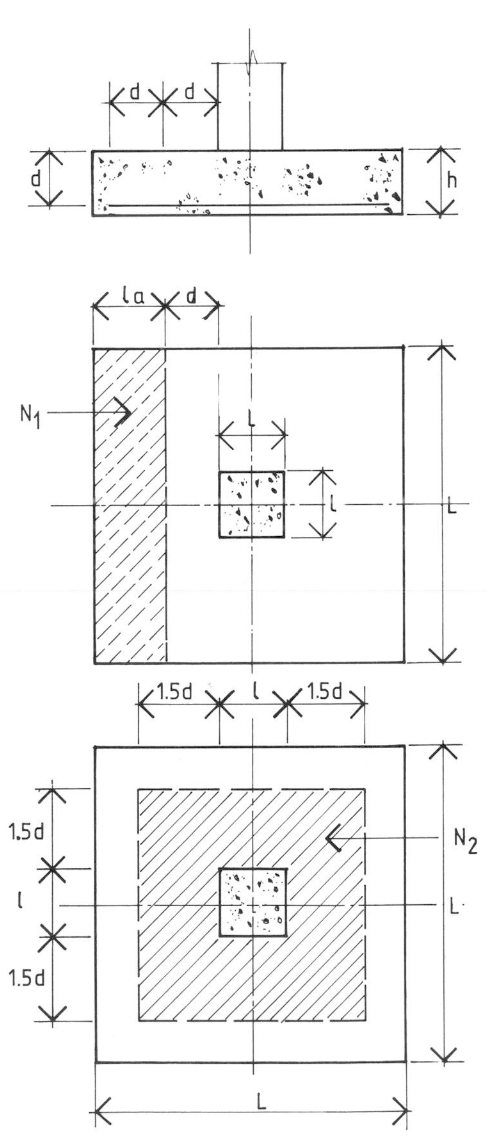

Figure 4.7 Column base

1. Convert 'design' column load to 'service' condition ($1{\cdot}0G_k + 1{\cdot}0Q_k$).
2. Assume a thickness of base and determine the weight per square metre (kN/m^2).
3. Determine the net allowable bearing pressure = q_s − weight of base (kN/m^2).
4. Calculate the area of base required = N_s/net allowable bearing pressure (m^2).
5. Determine side L(m).
6. Calculate $q_u = N_u/A$ (kN/m^2).
7. Choose the appropriate concrete cover from Table 4.13 and determine $d = h -$ (cover + 0·5 bar diameter).
8. Calculate $N_1 = q_u \times L \times L_1$ (kN/m^2).
9. Choose the material stresses to be used (if not already decided).
10. Determine $v_1 = N_1 - 0{\cdot}5N_d/L_1 d$ (N/mm^2).
11. Calculate $p = 41 + 12d$ (mm).
12. Calculate $A_1 = 12 + 61d + 9d^2$ (mm^2).
13. Determine $N_2 = q_u \times A_1$ (kN).
14. Determine $N_3 = N_u - N_2$ (kN).
15. Calculate $v = N_3/Pd$ (N/mm^2).
16. Calculate $M_u = N_u/8L(L - 1)^2$ (kNmm).
17. Determine the reinforcement area from $A_s = M_u/(0{\cdot}87f_y)z$.
18. Select the reinforcement required from Table 4.14 (same in both directions).
19. Determine $100A_s/bd$ and select the value of v_c from Table 4.12.
20. Check that $v_c > v_1$ and v_2. If not, redesign with increased base thickness.
21. Recommendations for the design of anchorage bond stress and local bond stress are given in BS 8100: Part 1.

4.9 Axially loaded columns

Introduction

A column may be defined as a compression member whose greater overall cross-sectional dimension does not exceed four times its smaller dimension. Where provisions relate primarily to rectangular cross sections, the principles involved may be applied to other shapes where appropriate.

Size of columns

The size of a column and the position of the reinforcement in it may be affected by the requirements for durability and fire resistance, and these should be considered before the design is commenced. However, the least dimension for a rectangular column should not be less than 200 mm.

Short and slender columns

A column may be considered as short when both the ratios l_{ex}/h and l_{ey}/b are less than 15 (braced) and 10 (unbraced) (Figure 4.8). It should otherwise be considered as slender.

Braced and unbraced columns

A column may be considered braced in a given plane if lateral stability to the structure as a whole is provided by a wall, bracing or buttressing designed to resist all lateral forces in that plane. It should otherwise be considered as unbraced.

Effective height of a column

The effective height l_e of a column in a given plane may be obtained from the following equation:

$$l_e = l_0$$

Values of β are given for braced columns in Table 4.18. BS 8100 should be consulted for a more rigorous assessment of the effectiveness of braced and unbraced columns.

It should be noted from Figure 4.8 that the effective height of a column in the two plan directions may be different.

Table 4.18 Column end conditions for braced columns

End condition at top	*End condition at bottom*		
	1	*2*	*3*
1	0·75	0·80	0·90
2	0·80	0·85	0·95
3	0·90	0·95	100

Values of β for braced columns
(a) *Condition 1.* The end of the column is connected monolithically to beams on either side which are at least as deep as the overall dimension of the column in the plane considered. Where the column is connected to a foundation structure, this should be of a form specifically designed to carry moment.
(b) *Condition 2.* The end of the column is connected monolithically to beams or slabs on either side which are shallower than the overall dimension of the column in the plane considered.
(c) *Condition 3.* The end of the column is connected to members which, while not specifically designed to provide restraint to rotation of the column will, nevertheless, provide some nominal restraint.

Axial loads and moments

The minimum design moment for any column in any plane should be obtained by multiplying the ultimate design axial load by an eccentricity which should be taken as 0·05 times the overall column dimension in the relevant plane, but not exceeding 20 mm. Alternatively, the moments in columns may be obtained using the methods outlined in BS 8100: Part 1, subject to the minimum design moments above.

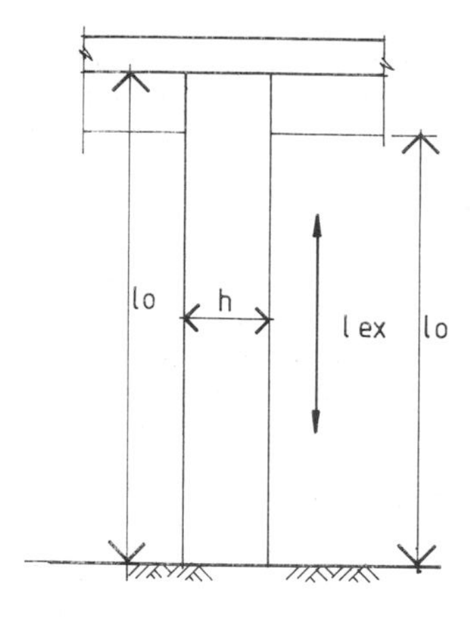

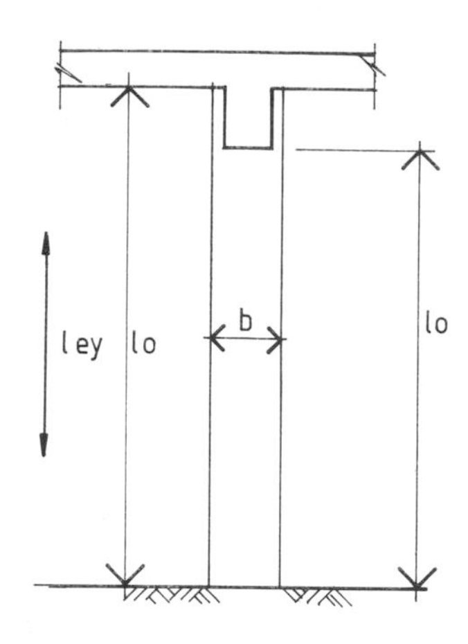

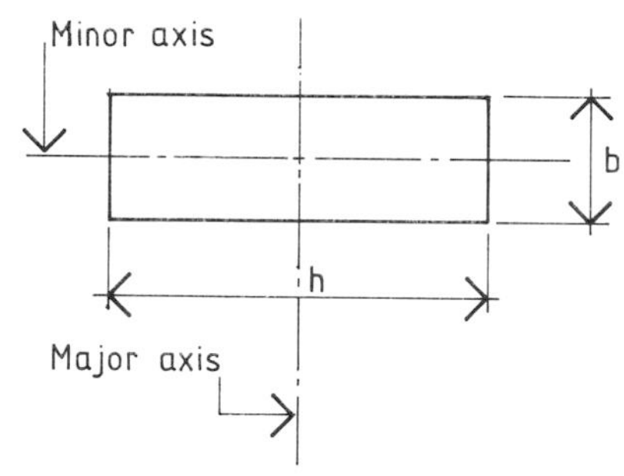

Figure 4.8 Effective column height

Basis of design

To find the amount of reinforcement in a column the following methods can be used:

1. Design charts.
2. Design formulae.

Design charts have been prepared using the stress–strain curves for concrete and steel as shown in BS 8100: Part 1. These charts, which are to be found in BS 8100: Part 3, are prepared for a particular grade of concrete and strength of reinforcement. A typical chart is shown in Figure 4.9.

Columns should normally be designed using the charts in BS 8100: Part 3 or one of the following simplified methods can be adopted:

1. In the case where, due to the nature of the structure, a column cannot be subjected to significant moments it may be designed so that the design ultimate axial load does not exceed the value of N given by:

   $N = 0{\cdot}4f_{cu}A_c + 0{\cdot}75f_yA_{sc}$

2. When a column supports an approximately symmetrical arrangement of beams (i.e. where adjacent spans do not differ by more than 15%) subject to uniformly distributed loads, the column may be designed so that the design ultimate axial load does not exceed the value of N given by:

   $N = 0{\cdot}35f_{cu}A_c + 0{\cdot}67f_yA_{sc}$

Design of an axially loaded column

1. Calculate the ultimate design axial load N on the column section.
2. Determine a cross section for the column (mm × mm) with a least dimension of 200 mm.
3. Check the dimensions are adequate for durability and fire resistance.
4. Determine the end column condition from Table 4.18 and calculate the effective height of the column.
5. Decide whether the column is braced or unbraced.
6. Calculate the slenderness ratios l_{ex}/h and l_{ey}/b for the column. If both ratios are less than 15 (braced) or 10 (unbraced) the column

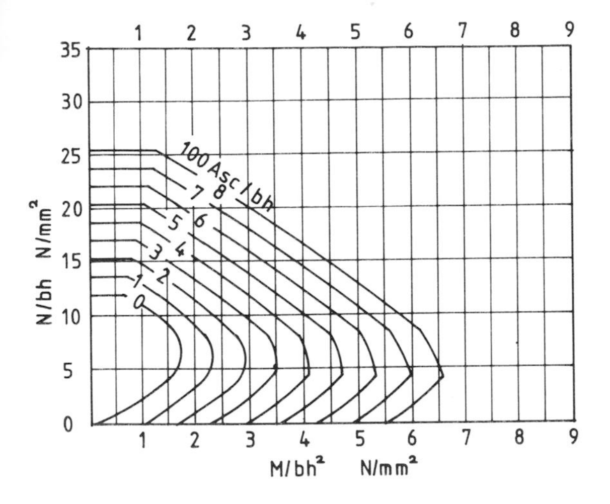

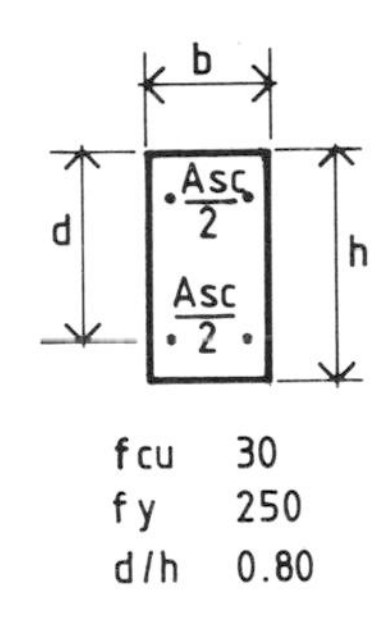

Figure 4.9 Typical design chart for columns

is considered as short; otherwise it is considered as slender and reference should be made to BS 8100: Part 1 for the method of design.

7. Decide on material stresses to be used, i.e. f_{cu} and f_y.
8. Calculate the reinforcement area by either:
   (a) Use of design charts: in which case, calculate N/bh and M/bh^2 and determine $100A_{sc}/bh$ from Figure 4.9 or the appropriate design chart in BS 8100: Part 3; or
   (b) Use of design formulae: in which case, calculate $A_{sc} = N - 0{\cdot}4f_{cu}A_c/0{\cdot}75f_y$ or $A_{sc} =$ N $- 0{\cdot}35f_{cu}A_c/0{\cdot}67f_y$, depending upon the loading condition on the column.
9. Select the reinforcement from Table 4.14.
10. Check the minimum and maximum reinforcement areas from BS 8100: Part 1, including the size and position of links.

5

Prestressed concrete

5.1 Introduction

The design procedures for all but the simplest of prestressed concrete members do not lend themselves, for the purpose of this book, to a simplified approach. The underlying structural principle of prestressing is in itself relatively simple. To illustrate this point an abbreviated design example for a simply supported beam of rectangular cross section will be used here.

In the context of structural concrete, prestressing is a term applied to those methods of construction where the flexural tension stresses due to dead and imposed loads are balanced, or significantly reduced by the application of a compressive force. This force is induced through high-strength steel or alloy tendons running through the length of the member. The advantage of this system over normal reinforced concrete is that the full compressive strength of the concrete is utilized under working conditions across most of the cross section of the member. The deflection of the member is also minimized. It is though, in comparison, a costly method of construction and tends therefore to be used in certain specialized cases only.

In specific terms there are two methods of achieving the prestressing effect: pre-tensioning and post-tensioning. In the case of pre-tensioning, the tendons are first stressed by the application of a tensioning force. The concrete is then cast around the tendons. When the concrete has achieved the required strength the tendons are severed at the ends of the member. As the tensile stress in the tendons is resisted by a compressive stress in the concrete through the bond between the tendons and the concrete this form of prestressing is often used in the production of factory-made units such as floor beams. With post-tensioning, ducts are cast in the concrete element through which the tendons are passed. After the concrete has reached a specified strength the tendons are stressed and anchored at their ends to the member, which has the effect of setting up a compressive stress in the concrete. This method is normally used for larger structures such as bridges, deck beams, etc. In both cases the force in the tendons is produced by using a hydraulic jack. When this force is applied to the concrete it is termed the 'transfer of stress'. For pre-tensioned members this takes place when the tendons are severed. In post-tensioned work it occurs as soon as the tendons are stressed.

5.2 Design requirements from BS 8110, *The structural use of concrete*

Concrete

In order to take full advantage of prestressing, high-strength concrete should be used. The minimum characteristic strengths recommended are 30 N/mm^2 and 40 N/mm^2 for post-tensioning and pre-tensioning, respectively. When prestressed concrete members are in service the compressive strength of the concrete should not exceed the values given in Table 5.1. The allowable compressive strength of the concrete at transfer of stress is usually greater than the concrete strength in service (see also 'Loss of prestress' below). Values of allowable compressive strengths at transfer are given in Table 5.2. In some casees tensile stresses in flexure are permitted. Reference, however, should be made to the relevant section of BS 8110: Part 1.

Table 5.1 Compressive stresses for concrete in service (f_{cs})

Nature of loading	*Allowable compressive stresses*
Design load in bending	$0{\cdot}33\ f_{cu}$ (for simply supported beams)
Design load in direct compression	$0{\cdot}25 f_{cu}$

f_{cu}: Characteristic concrete cube strength.

Table 5.2 Allowable compressive stresses at transfer f_{ci}

Nature of stress distribution	*Allowable compressive stress*
Triangular or near-triangular distribution of prestress	$0{\cdot}5 f_{ci}$
Near-uniform distribution of prestress	$0{\cdot}4 f_{ci}$

f_{ci} is the concrete strength at transfer.

Steel

The steel used in prestressing work is usually in the form of high-tensile wires or alloy steel bars.

Definition of terms

Tendon. A stretched element used in a concrete member to impart prestress to the concrete. It can consist of individual steel wires, bars or strands.
Cable. A group of tendons.
Wire. Reinforcement of solid cross section up to 7 mm in diameter complying with the requirements of BS 2691: 1969.
Strands. A group of either seven or 19 wires wound in helical form complying with the requirements of BS 3617: 1971 and BS 4757: 1971.
Bars. Reinforcement of solid cross section up to 40 mm in diameter complying with the requirements of BS 4456: 1969.

Values of characteristic strength for wires, strands or bars are given in Tables 5.3–5.5.

Table 5.3 Specified characteristic strengths of prestressing wire

Nominal size (mm)	*Specified characteristic strength (kN)*	*Nominal cross-sectional area (mm^2)*
2	6·34	3·14
2·65	10·3	5·5
3	12·2	7·1
3·25	14·3	8·3
4	21·7	12·6
4·5	25·7	15·9
5	30·8	19·6
7	60·4	38·5

Table 5.4 Specified characteristic strengths of prestressing strands

Number of wires	*Nominal size (mm)*	*Specified characteristic strength (kN)*	*Nominal cross-sectional area (mm^2)*
7	6·4	44·5	24·5
	7·9	69·0	37.4
	9·3	93·5	52·3
	10·9	125	71·0
	12·5	165	94·2
	15·2	227	138·7
19	18	370	210
	25·4	659	423
	28·6	823	535
	31·8	979	660

Table 5.5 Specified characteristic strengths of prestressing bars

Nominal size (mm)	*Specified characteristic strength (kN)*	*Nominal cross-sectional area (mm^2)*
20[a]	325	314
22	375	380
25[a]	500	491
28	625	615
32[a]	800	804
35	950	961
40[a]	1250	1257

[a] Preferred sizes.

Loss of prestress

There are a number of ways in which the initial prestressing force exerted by the jack is not effectively retained after transfer. This is known as loss of prestress. The causes can be summarized as follows:

1. Shrinkage of concrete.
2. Elastic deformation of concrete.
3. Creep of concrete.
4. Relaxation of steel.
5. Steam curing.
6. Friction.

Maximum initial prestress

The jacking force should not normally exceed 75% of the characteristic strength of the tendon. In some cases the force can be increased to 80% but reference should be made to BS 8110: Part 1. It should be noted that the initial prestress at transfer should not exceed 70% of the characteristic strength of the tendons and in no case should it exceed 75%.

Cover

General rules for the cover to prestressing tendons are given in BS 8110: Part 1.

Effective span

The effective span of a simply supported member should be taken as the smaller of:

1. The distance between the centres of bearings, or
2. The clear distance between supports, plus the effective depth.

5.3 Design of a rectangular simply supported pre-tensioned beam (no tension develops)

Symbols

A	cross-sectional area of beam (mm^2)
e	eccentricity of prestressing force (mm)
f_{ci}	allowable compressive stress for concrete at transfer (N/mm^2)
f_{cs}	allowable compressive stress for concrete in service (N/mm^2)
M_d	bending moment due to deadweight of beam (N mm)
M_l	bending moment due to additional dead and live loads (N mm)
M_s	total service moment at any section $M_d + M_l$ (N mm)
P	prestressing force in tension (N).
Z	section modulus (mm^3)
α	$\dfrac{\text{Effective force in tendon after losses}}{\text{Force in tendon at transfer}}$

Procedure for preliminary calculations to determine the minimum required prestress force *p*(min) and the maximum eccentricity *e*(max) for a Class 1 rectangular simply supported prestressed concrete beam

1. Calculate the maximum bending moment due to the superimposed live load:

   $$M_l = \frac{wl^2}{8} \text{ (N mm)}$$

   where w = the superimposed live load per metre run of beam and l = the effective span.
2. Assume a value for the maximum bending moment due to the self-weight of the beam, i.e.

   35% of M_l

   Hence $M_d = 0{\cdot}35M_l$ (N mm).
3. Calculate the total service moment

   $$M_s = M_d + M_l$$
4. Calculate the minimum section modulus Z_t required to keep within the allowable compression stress in the top fibre of the concrete in service:

   $$Z_t = \frac{M_s}{f_{cs}}$$

   For values of f_{cs} see Table 5.1.
5. Calculate the minimum section modulus Z_b required to keep within the allowable compression stress in the bottom fibre of the concrete due to the initial prestress:

   $$Z_b = \frac{M_s}{\alpha f_{ci}}$$

   For α assume a value of 0·8 and for values of f_{ci} see Table 5.2.
6. A section for the beam must now be chosen based on the requirements of Z_t and Z_b. It is also necessary to check that the actual dead load moment M_d is less than or equal to the assumed value of $0{\cdot}35M_l$. If the section chosen satisfies these conditions then:
7. Calculate the minimum required prestressing force

   $$P(\text{min}) = \frac{(M_s - \alpha M_d)}{\alpha(Z_t + Z_b)} A \quad \text{(N)}$$
8. Calculate the maximum allowable eccentricity of the prestressing force:

   $$e(\text{max}) = \frac{Z_b}{A} + \frac{M_d}{P(\text{min})} \quad \text{(mm)}$$

Figure 5.1 shows the critical stress distributions in the beam at the moment of transfer, and Figure 5.2 shows the critical stress distributions in the beam under service conditions.

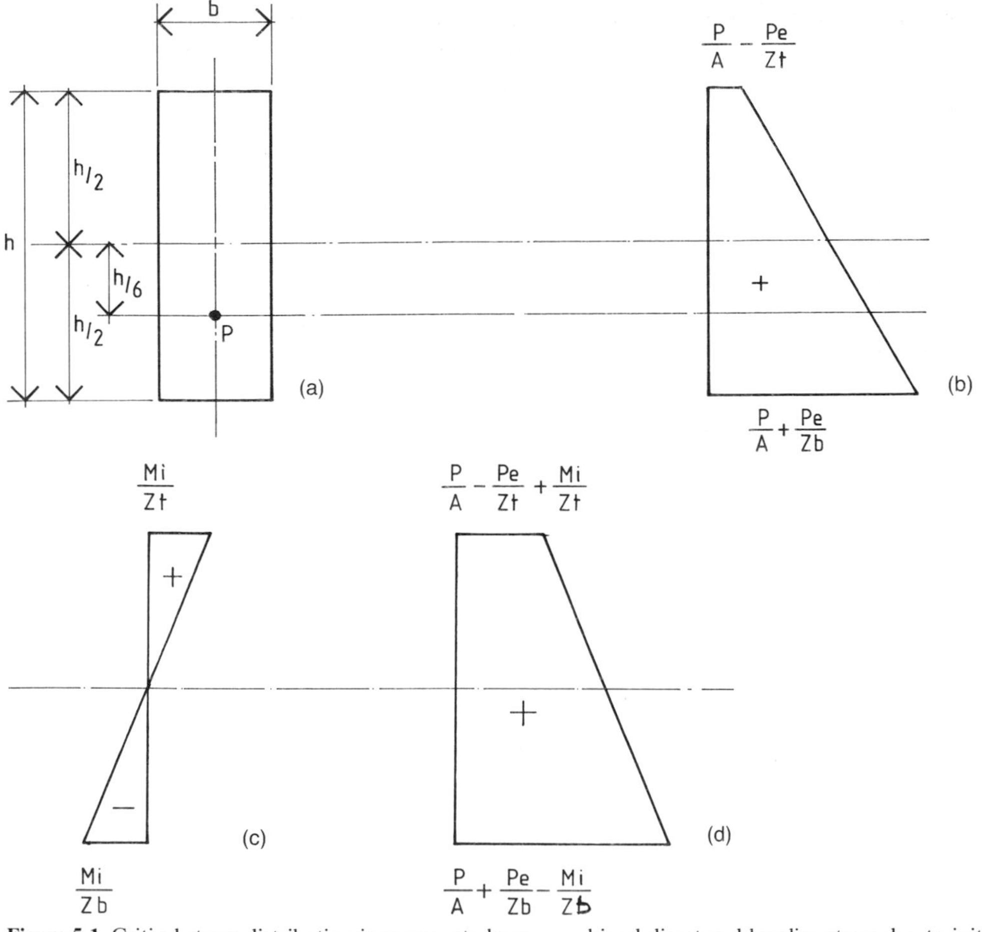

Figure 5.1 Critical stress distribution in a concrete beam at moment of transfer. (a) Cross section through beam showing assumed position of cable force P; (b) combined direct and bending stress due to initial prestress; (c) bending stress due to initial moment; (d) transfer (initial) condition

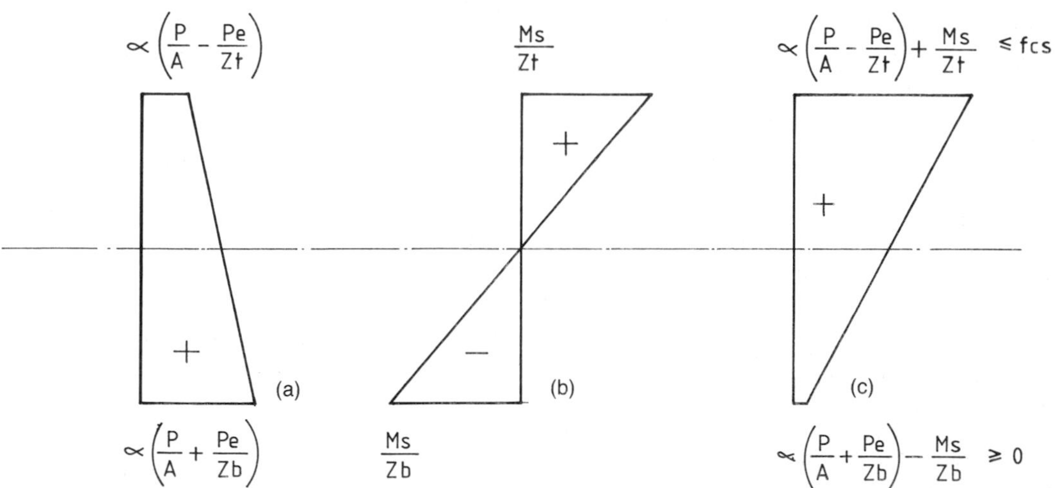

Figure 5.2 Critical stress distribution in a concrete beam under service conditions. (a) Combined direct and bending stress due to prestress in service; (b) bending stress due to service moment; (c) stress condition in service

It is advisable to check with the calculated values of P(min) and e(max) that the above conditions are satisfied.

It should be noted that these results are only preliminary, and further calculations would be required to determine the actual loss of prestress (an assumption has been made at this stage). The requirements of ultimate limit state and deflection should also be satisfied.

6

Loadbearing brickwork and blockwork

6.1 Design requirements

BS 5628: Parts 1–3 gives recommendations for the design of structural brickwork and blockwork as well as the requirements of good workmanship.

Limit state

This can be defined as the achievement of acceptable probabilities that the brick or block wall being designed will not become unfit for the use for which it is required.

Characteristic loads

Since it is not, at present, possible to express loads in statistical terms, the following characteristic loads should be included in the design:

1. Dead loads G_k. The weight of the structure complete with finishes, partitions, etc.
2. Imposed loads Q_k. Weight due to furniture, occupants, etc.
3. Wind loads W_k.

The following publications should be used to compute the loads on the structure:

BS 648: 1964, *Schedule of Weights of Building Materials*.
BS 6399: Part 1, *Dead and Imposed loads*.
CP3 : Chapter V: Part 2: 1972 *Wind Loads*.

Characteristic strength of masonry

The value of the strength of masonry below which the probability of test results failing is not more than 5%. The characteristic compressive strength may be based on the results of laboratory tests or on information given in BS 5628: Part 1: 1978. Typical values of characteristic strength are given in Tables 6.1–6.3.

It should be noted that where the horizontal cross-sectional area of a loaded wall or column is less than $0{\cdot}2\ m^2$ the characteristic compressive strength should be multiplied by the factor $(0{\cdot}70 + 1{\cdot}5A)$, where A = horizontal loaded cross-sectional area of the wall or column.

Partial safety factors

Loads

In this case partial safety factors are introduced to take account of unforeseen variations in the characteristic loads. These factors differ for dead, imposed and wind loads and are set out in Table 6.4.

Materials

Such safety factors are introduced to allow for possible differences between the characteristic strength and the actual structural strength. The value of these factors are generally taken as shown

Table 6.1 Characterisitc compressive strength of masonry f_k constructed with standard-format bricks

Mortar designation	*Compressive strength of unit (N/mm²)*								
	5	*10*	*15*	*20*	*27.5*	*35*	*50*	*70*	*100*
(i)	2·5	4·4	6·0	7·4	9·2	11·4	15·0	19·2	24·0
(ii)	2·5	4·2	5·3	6·4	7·9	9·4	12·2	15·1	18·3
(iii)	2·5	4·1	5·0	5·8	7·1	8·5	10·6	13·1	15·5
(iv)	2·2	3·5	4·4	5·2	6·2	7·3	9·0	10·8	12·6

Table 6.2 Characteristic compressive strength of solid concrete blocks f_k having a ratio of height to least horizontal dimension of between 2·0 and 4·0

Mortar designation	*Compressive strength of unit (N/mm²)*							
	2·8	*3·5*	*5·0*	*7·0*	*10*	*15*	*20*	*35 or greater*
(i)	2·8	3·5	5·0	6·8	8·8	12·0	14·8	22·8
(ii)	2·8	3·5	5·0	6·4	8·4	10·6	12·8	18·8
(iii)	2·8	3·5	5·0	6·4	8·2	10·0	11·6	17·0
(iv)	2·8	3·5	4·4	5·6	7·0	8·8	10·4	14·6

Table 6.3 Mortar mixes and strengths

Mortar designation	*Type of mortar (proportion by volume)*			*Mean compressive strength at 28 days*	
	Cement:lime: sand	*Masonry cement: sand*	*Cement:sand with plasticizer*	*Preliminary (laboratory) tests (N/mm²)*	*Site tests (N/mm²)*
(i)	1:0 to ¼:3	–	–	16·0	11·0
(ii)	1:½:4 to 4½	1:2½ to 3½	1:3 to 4	6·5	4·5
(iii)	1:1:5 to 6	1:4 to 5	1:5 to 6	3·6	2·5
(iv)	1:2:8 to 9	1:5½ to 6½	1:7 to 8	1·5	1·0

Table 6.4 Partial safety factor y_f for design loads

(a) Design and imposed load

Loading	*Partial safety factor*
Design dead load	$0{\cdot}9G_k$ or $1{\cdot}4G_k$
Design imposed load	$1{\cdot}6Qk$

(b) Dead and wind load

Loading	*Partial safety factor*
Design dead load	$0{\cdot}9G_k$ or $1{\cdot}4G_k$
Design wind load	$1{\cdot}4W_k$ or $0{\cdot}015G_k$ whichever is the greater

(c) Dead, imposed and wind load

Loading	*Partial safety factor*
Design dead load	$1{\cdot}2G_k$
Design imposed load	$1{\cdot}2Q_k$
Design wind load	$1{\cdot}2W_k$ or $0{\cdot}015G_k$ whichever is the greater

(d) Accidental damage

Loading	*Partial safety factor*
Design dead load	$0{\cdot}095G_k$ or $1{\cdot}05G_k$
Design imposed load	$0{\cdot}35Q_k$ (except that in the case of buildings used predominantly for storage, or where the imposed load is of a permanent nature, $1{\cdot}05Q_k$ should be used)
Design wind load	$0{\cdot}35W_k$

Notes: 1. Where alternative values are shown, that producing the most severe conditions should be selected.
2. In design, each of the load combinations (a)–(d) should be considered and that giving the most severe conditions should be adopted.
3. G_k: characteristic dead load.
Q_k: characteristic imposed load.
W_k: characteristic wind load.

in Table 6.5, and allow for the difference between laboratory- and site-constructed brickwork and blockwork.

Table 6.5 Partial safety factors γ_m for material strength

		Category of construction control	
Category of manufacturing control of structural units	Special	2·5	3·1
	Normal	2·8	3·5

Note: For a description of categories of manufacturing and construction control see BS 5628: Part 1: 1978.

Design loads

These can be defined as:

Characteristic load × partial safety factor

As previously mentioned, the partial safety factors vary according to the circumstances under which the loads are considered. Maximum design loads can be obtained by reference to Table 6.4.

Effective height

The effective height of a wall may be taken as:

1. 0·75 × the clear distance between lateral supports which provide enhanced resistance to lateral movement, or
2. The clear distance between lateral supports which provide simple resistance to lateral movement.

The effective height of a column should be taken as:

1. The distance between lateral supports, or
2. Twice that distance in respect of a direction in which lateral support is not provided.

The effective height of a pier may be treated as a wall for effective height consideration if the thickness of the pier is not greater than 1·5 times the thickness of the wall of which it forms a part. The pier should otherwise be treated as a column in the plane at right angles to the wall. The terms 'enhanced' and 'simple' resistance are defined in BS 5628: Part 1: 1978. Examples are given in Figures 6.1 and 6.2.

Effective length

The effective length of a wall may be taken as:

1. 0·75 × the clear distance between vertical lateral supports or

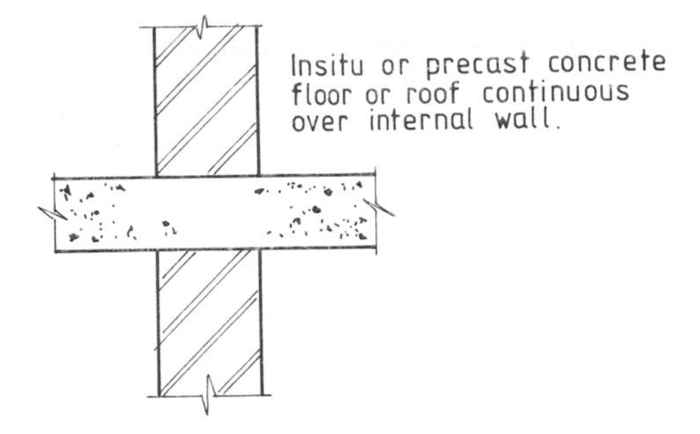

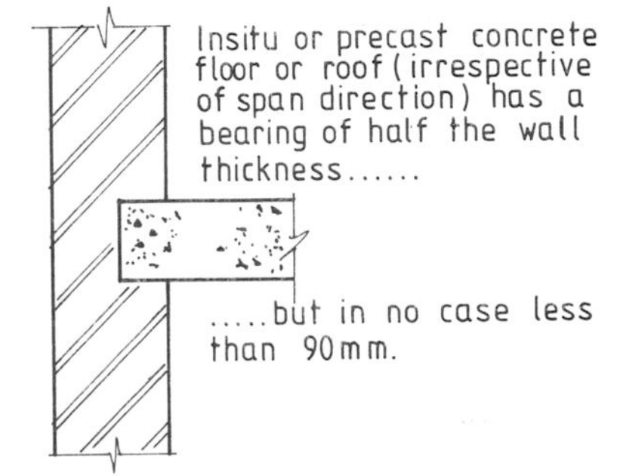

Figure 6.1 Examples of enhanced resistance to lateral movement of wall

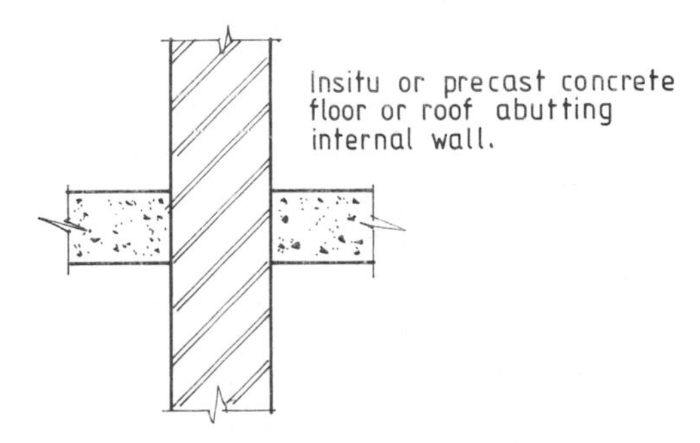

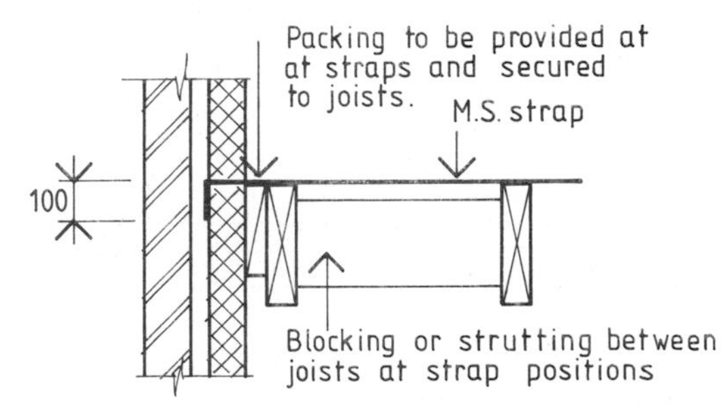

Figure 6.2 Examples of simple resistance to lateral movement of wall (other examples of simple resistance are shown in BS 5628: Part 1)

2. Twice the distance between a support and a free edge, where lateral supports provide enhanced resistance to lateral movement; or
3. The clear distance between lateral supports; or
4. 2·5 × the distance between a support and a free edge, where lateral supports provide simple resistance to lateral movement.

Effective thickness

The effective thickness of a single-leaf wall or column is the actual thickness. With a cavity wall or column this thickness should be taken as:

1. $\frac{2}{3}(t_1 + t_2)$, or
2. t_1, or
3. t_2,

where t_1 and t_2 = actual thickness of the brick or block leaf.

When a wall is stiffened by piers or intersecting walls the effective thickness of a single-leaf wall (which may be one leaf or a cavity wall) is given by:

$t_{ef} = t \times K$

where:

t_{ef} = the effective thickness of the wall or leaf,
t = the actual thickness of the wall or leaf,
K = the appropriate stiffness coefficient taken from Table 6.6.

If intersecting walls are used to stiffen a wall instead of piers, assume the intersecting wall to be a pier, where b = thickness of the intersecting wall and $t_p = 3 \times$ the thickness of the stiffened wall. The appropriate stiffness coefficient may then be obtained from Table 6.6.

Table 6.6 Stiffness coefficient *K* for walls stiffened by piers

Ratio of pier spacing (centre to centre) to pier width	*Ratio of pier thickness to actual thickness of wall to which it is bonded*		
	1	*2*	*3*
6	1·0	1·4	2·0
10	1·0	1·2	1·4
20	1·0	1·0	1·0

Slenderness ratio

This can be defined as:

$$\frac{\text{Effective height}}{\text{Effective thickness}}$$

BS 5628: Part 1: 1978 recommends that for walls set in Portland cement mortars the slenderness ratio should not exceed 20 for walls less than 90 mm thick in buildings of more than two storeys and 27 in all other cases. Table 6.7 gives reduction factors for slenderness ratios 0 to 27 for varying eccentricities at the top of the wall and Figure 6.3 illustrates various assessments of eccentricities.

Table 6.7 Capacity reduction factor β

Slenderness ratio	*Eccentricity at top of wall*			
	Up to 0·05t	*0·1t*	*0·2t*	*0·3t*
0	1·00	0·88	0·66	0·44
6	1·00	0·88	0·66	0·44
8	1·00	0·88	0·66	0.44
10	0·97	0·88	0·66	0·44
12	0·93	0·87	0·66	0·44
14	0·89	0·83	0·66	0·44
16	0·83	0·77	0·64	0·44
18	0·77	0·70	0·57	0·44
20	0·70	0·64	0·51	0·37
22	0·62	0·56	0·43	0·30
24	0·53	0·47	0·34	
26	0·45	0·38		
27	0·40	0·33		

Note: It is not necessary to consider the effects of eccentricities up to less than or equal to 0·5*t*.

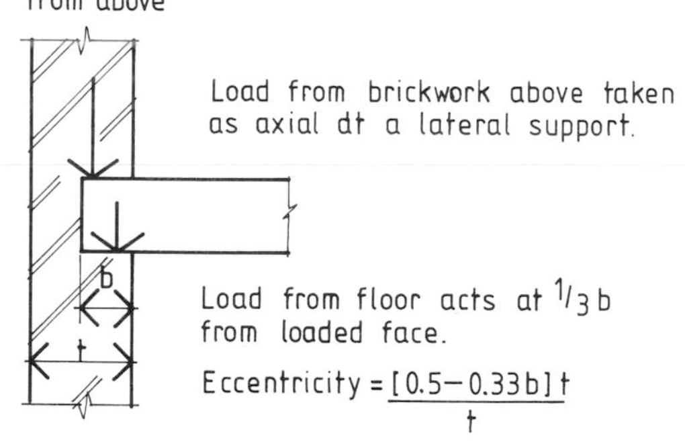

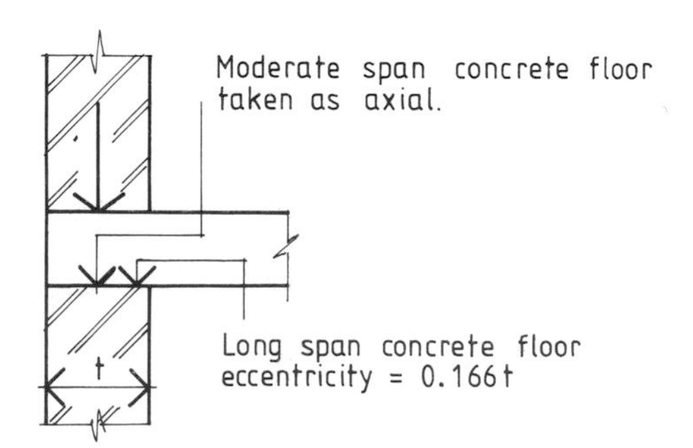

Figure 6.3 Typical examples of assessment of eccentricity

Design vertical load resistance of walls

This is given by

$$\frac{\beta t f_k}{\gamma_m} \text{ per unit length of wall}$$

where

β = capacity reduction factor allowing for the effects of slenderness and eccentricity (see Table 6.7),

f_k = characteristic strength of the masonry obtained from Tables 6.1 or 6.2,

t = the thickness of the wall,

γ_m = the partial safety factor for the material obtained from Table 6.5.

Design vertical load resistance of columns

This is given by

$$\frac{\beta b t f_k}{\gamma_m}$$

where

f_k = characteristic strength of the masonry obtained from Tables 6.1 or 6.2,

t = the thickness of column,

b = width of column,

β = capacity reduction factor determined in accordance with the recommendations of BS 5628: Part 1: 1978.

Design vertical load resistance of cavity walls and columns

When the applied vertical load acts between the centroids of the two leaves of a cavity wall or walls it should be replaced by two equivalent axial loads on the two leaves, and then designed in accordance with the above.

Concentrated loads (design stress reductions)

BS 5628: Part 1: 1978 suggests three types of concentrated load application where the normal design stress may be reduced.

Type 1. The stresses obtained from concentrated loads bearing over substantial areas of the wall (Figure 6.4) may be increased by 1·25 × the normal design stress.

Type 2. The stresses obtained from concentrated loads bearing over very limited areas of the wall (Figure 6.5) may be increased by 1·50 × the normal design stress.

Type 3. The stresses obtained from concentrated loads borne by properly designed spreader beams located at the end of the wall (Figure 6.6) may be increased by 2·0 × the normal design stress.

It should be remembered that all other loadings on the same areas as the concentrated load must be taken into account in the design.

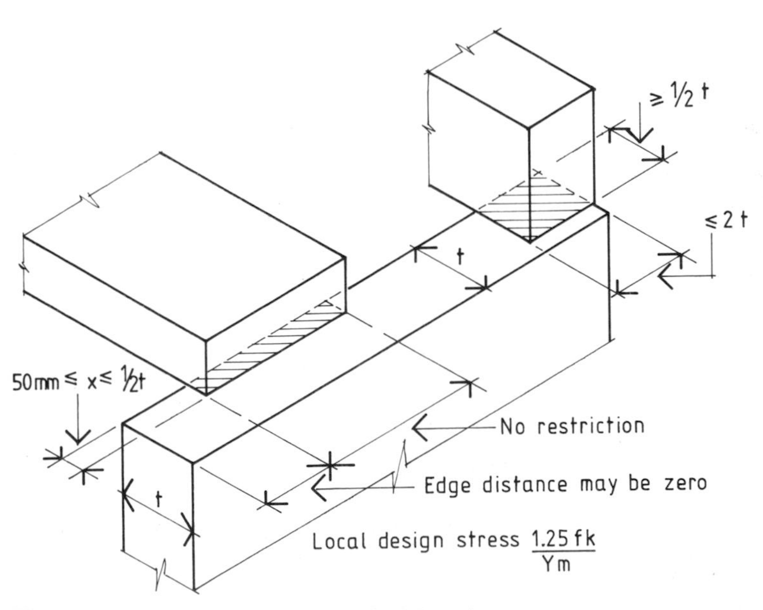

Figure 6.4 A typical concentrated load of Type 1

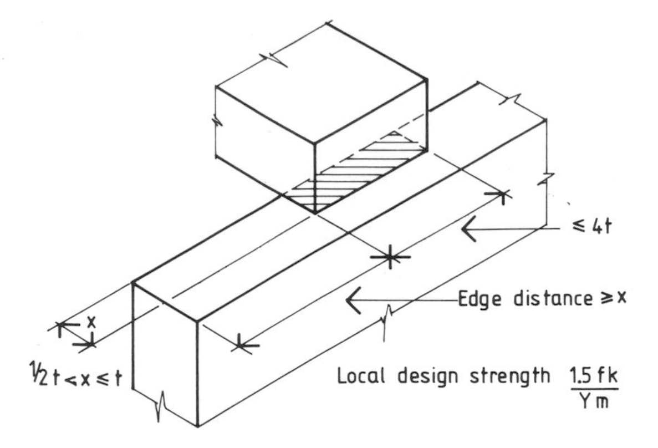

Figure 6.5 A typical concentrated load of Type 2

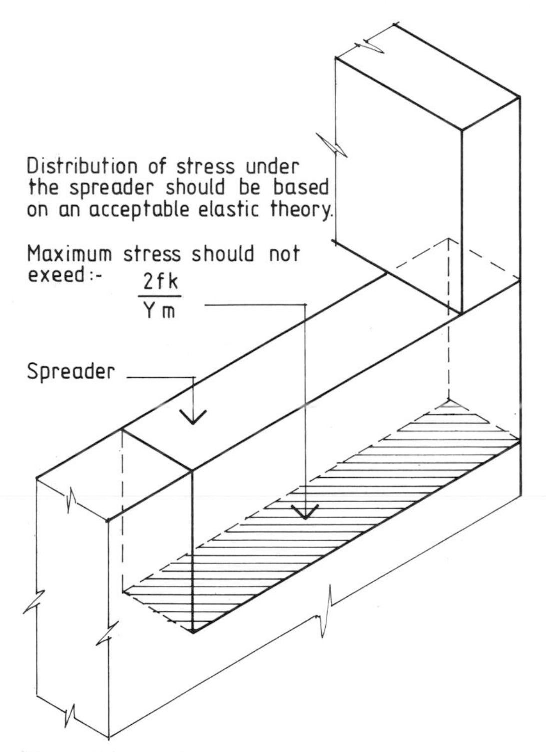

Figure 6.6 A typical concentrated load of Type 3

Walls subjected to shear forces

BS 5628: Part 1: 1978 should be consulted where walls resist, in shear, horizontal forces acting in their plane.

6.2 Examples of determining the effective thickness of brick and block walls

1. Determine the effective thickness for the unstiffened cavity wall shown in Figure 6.7.

Case (a) $t_{ef} = \frac{2}{3}(t_1 + t_2)$
$t_{ef} = \frac{2}{3}(150 + 75) = 150$ mm
Case (b) $t_{ef} = t_1 = 150$ mm
Case (c) $t_{ef} = t_2 = 75$ mm

In this example the effective thickness of the wall = 150 mm.

2. Determine the effective thickness for the stiffened single-leaf wall shown in Figure 6.8.

$$\frac{\text{Pier spacing}}{\text{Pier width}} = \frac{4400}{440} = 10$$

$$\frac{\text{Pier thickness}}{\text{Actual wall}} = \frac{430}{215} = 2$$

From Table 6.6 the stiffness coefficient $K = 1{\cdot}2$. Hence

$t_{ef} = 1{\cdot}2t$

$t_{ef} = 1{\cdot}2 \times 215 = 258$ mm

Effective thickness of wall = 258 mm

3. Determine the effective thickness for the stiffened cavity wall shown in Figure 6.9.

$$\frac{\text{Pier spacing}}{\text{Pier width}} = \frac{4400}{440} = 10$$

$$\frac{\text{Pier thickness}}{\text{Actual wall thickness}} = \frac{215}{102{\cdot}5} = 2{\cdot}09$$

From Table 6.6 stiffness coefficient (K) = 1·22. Hence

$t_{1ef} = 1{\cdot}22 \times 102{\cdot}5$

$= 125{\cdot}1$ mm

Therefore $t_{ef} = \frac{2}{3}(125{\cdot}1 + 102{\cdot}5)$

$t_{ef} = 152{\cdot}73$ mm

In this example the effective thickness of the wall = 151·73 mm

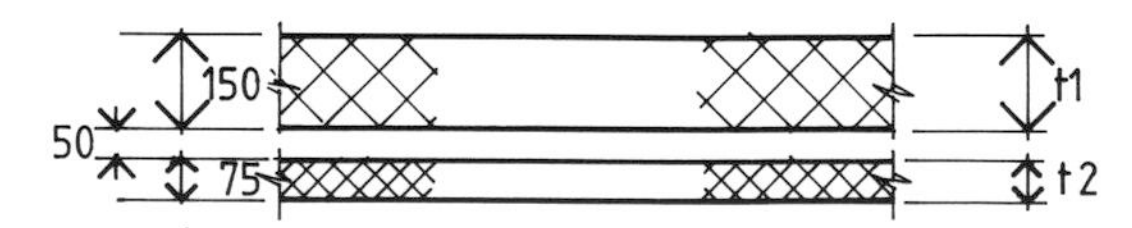

Figure 6.7 An unstiffened cavity wall

6.3 Design of a simple single-leaf wall

1. Estimate the characteristic loads G_k and Q_k (kN).

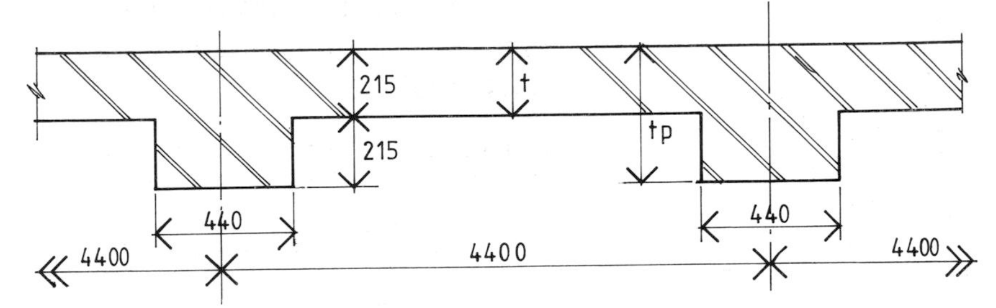

Figure 6.8 A stiffened single-leaf wall

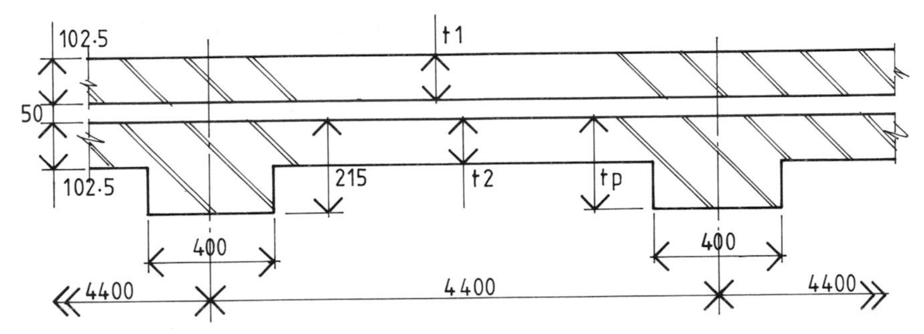

Figure 6.9 A stiffened cavity wall

2. Calculate the design loads using partial safety factors from Table 6.4.
3. Assess the type of lateral support at the top and bottom of the wall and determine the effective height (mm).
4. Calculate the effective thickness of the wall if necessary (mm).
5. Calculate the slenderness ratio of the wall from:

$$\frac{\text{Effective height}}{\text{Effective thickness}}$$

6. Assess the degree of eccentricity of loading at the top of the wall.
7. Select the capacity reduction factor from Table 6.7.
8. Select the partial safety factor for the material strength y_m from Table 6.5.
9. Calculate the characteristic compressive strength of the masonry from:

$$f_k = \frac{\text{Design load} \times y_m}{\beta \times t}$$

10. Select the appropriate mortar designation and type of mortar from Table 6.3.
11. Select the brick or block strength from Tables 6.1 or 6.2.

7

Retaining walls

For earth-retaining structures the relevant code is Civil Engineering Code of Practice No. 2 (1951). It should be noted that some of the information contained in this publication is still in imperial units. It is, however, anticipated that the code will be revised and published as a British Standard at a future date.

7.1 Pressures on retaining walls due to soil and other granular materials

The pressures considered in this book apply to cohesionless soils such as sands and gravels. Other materials which possess the property of internal friction (for example, coal, coke grain and certain ores) may also be considered under this heading. Pressures due to cohesive soils such as silts and clays are too complex to be dealt with in a work of this description and reference should be made to the code.

Symbols used in calculating pressures on retaining walls

B breadth of the base or foundation of the wall
D depth of the foot of foundation below ground level at front of wall
d distance of a point below the earth surface in front of the wall
e eccentricity of the resultant thrust
f_1 pressure on the foundation soil at the toe of the wall
f_2 pressure on the foundation soil at the heel of the wall
H vertical height of the earth retained by the wall and foundation
h net vertical height of the earth retained by the wall
h_w vertical distance to ground water level
K_a coefficient of active earth pressure for cohesionless soils
L length generally
P_a total active lateral thrust per unit length of wall on the wall back due to earth alone
p_a intensity of active lateral pressure at a given depth due to earth alone
P_p total passive resistance of earth in front of the wall
p_p intensity of the passive resistance of the earth in front of the wall at a given depth below the surface
P_u total lateral thrust per unit length of wall on wall back due to uniformly distributed load
p_u intensity of lateral pressure at a given depth

due to a superimposed uniformly distributed surcharge
p_w intensity of water pressure on the wall back at a given depth
q maximum safe bearing capacity of the soil
R resultant thrust on the foundation soil per unit length of wall
R_H horizontal component of the resultant thrust on the foundation soil per unit length of wall
R_v vertical component of the resultant thrust on the foundation soil per unit length of wall
W weight per unit length
W_s intensity of surcharge loading per unit area
y_a average density of all the strata down to a given depth
y_b submerged density of soil
y_d dry density of soil
y_m moist density of soil
y_s saturated density of soil
y_w density of water
z vertical distance measured behind the wall
δ angle of friction between the retained earth and wall back
ϕ angle of internal friction of the retained earth

7.2 Active pressures on a vertical wall with cohesionless soil backing

Horizontal ground (Figures 7.1 and 7.2)

The active pressure intensity at any depth below the horizontal ground surface is given by the equation

$$p_a = K_a y z \sec\delta \text{ (kN/m}^2)$$

The total active pressure

$$P_a = K_a y \frac{H^2}{2} \sec\delta \text{ (kN/m)}$$

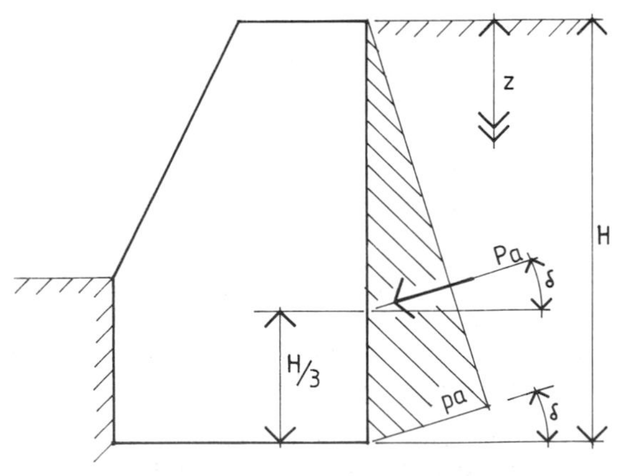

Figure 7.1 Active pressure diagram (general case)

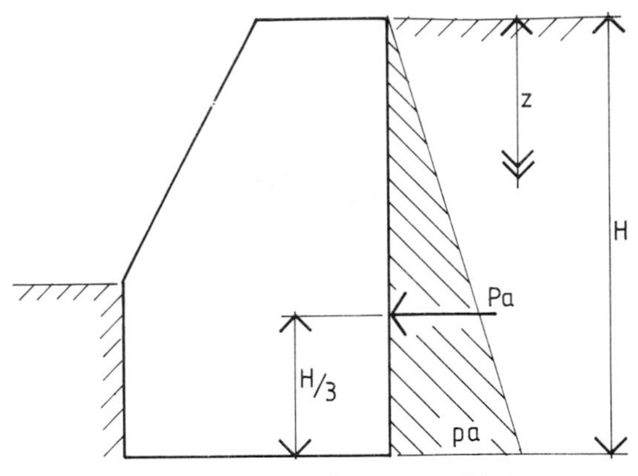

Figure 7.2 Active pressure diagram (Rankine)

Rankine's formula assumes that $\delta = 0$: therefore $\sec 0 = 1$ and hence

$$p_a = K_a y z \text{ (kN/m}^2)$$

$$p_a = K_a y \frac{H^2}{2} \text{ (kN/m)}$$

$$K_a = \frac{(1 - \sin\phi)}{(1 + \sin\phi)} = \tan^2\left(45 - \frac{\phi}{2}\right)$$

Typical values of ϕ and K_a are shown in Tables 7.1 and 7.2 and densities of cohesionless materials are given in Table 7.3.

Table 7.1 Typical values of ϕ for cohensionless materials

Materials	ϕ
Sandy gravel	35–45°
Compact sand	35–40°
Loose sand	30–35°
Shale filling	30–35°
Rock filling	35–45°
Ashes or broken brick	35–45°

Table 7.2 Values of K_a for cohesionless materials (vertical walls and horizontal ground)

Values of δ	*Values of φ* 25°	30°	35°	40°	45°
	Values of K_a				
0°	0·41	0·33	0·27	0·22	0·17
10°	0·37	0·31	0·25	0·20	0·16
20°	0·34	0·28	0·23	0·19	0·15
30°	–	0·26	0·21	0·17	0·14

Table 7.3 Densities of cohesionless materials (kg/m³)

Material		*Density when drained above ground water level (y_m)*	*Density when submerged below ground water level (y_b)*
Gravel		1600–2000	960–1280
Coarse and medium sands		1680–2080	
Fine and silty sands		1760–2160	
Rocks	granites and shales	1600–2080	960–1280
	basalts and dolerites	1760–2240	1120–1600
	limestones and sandstones	1280–1920	640–1280
	chalk	960–1280	320–640
Broken brick		1120–1760	640–960
Ashes		640–960	320–480

Fully saturated density = submerged density + 1000 kg/m³
$y_s = y_b + y_w$

Superimposed uniformly distributed load: horizontal ground (Figure 7.3)

Rankine replaced the superimposed uniformly distributed load by an equivalent height of earth: hence:

$$h_e = \frac{W_s}{y}$$

where

h_e = equivalent height of earth,

W_s = superimposed UDL,

y = density of material.

Therefore the active pressure intensity at any depth below the horizontal ground surface is given by the equation

$$p_u = K_a y h_e \text{ (kN/m}^2\text{)}$$

The total active pressure

$$P_u = K_{ay} H_e H \text{ (kN/m)}$$

Ground sloping at an angle equal to the angle of repose (Figure 7.4)

In this case Rankine assumed that the angle of repose is equal to the angle of internal friction. Therefore the active pressure intensity at any depth is given by the equation

$$p_a = yz \cos \phi \text{ (kN/m}^2\text{)}$$

The total active pressure

$$P_a = y \frac{H^2}{2} \cos \phi \text{ (kN/m)}$$

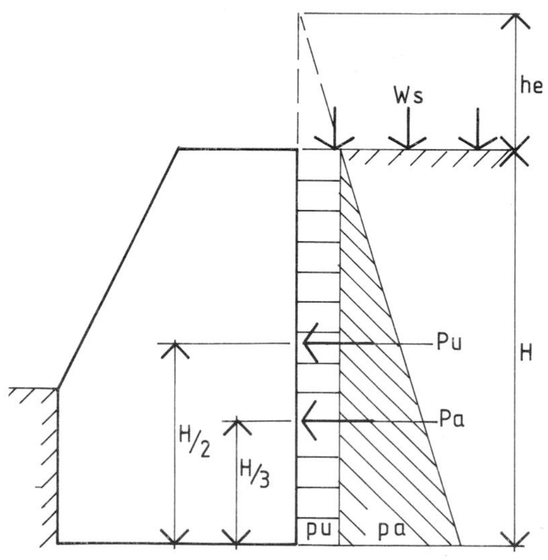

Figure 7.3 Active pressure and surcharge diagram

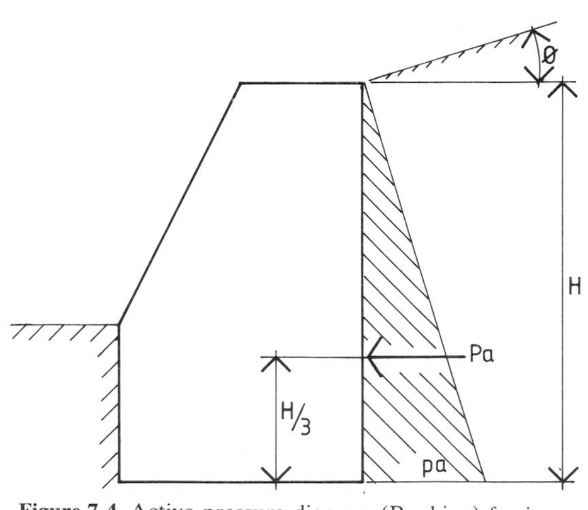

Figure 7.4 Active pressure diagram (Rankine) for inclined ground

However, another method which the student may consider for this case is Coulomb's wedge theory. Space does not allow a description here of the wedge theory but it is fully described in most books on the mechanics of engineering soils.

Saturated soils: horizontal ground (Figure 7.5)

If the ground water level is at a depth h_w below the surface of the ground the backing will be waterlogged below this level and its density will be the submerged value as shown in Table 7.3. Rankine's formula then becomes

$$p_1 = K_a y_m h_w \text{ (kN/m}^2\text{)}$$

$$P_1 = K_a y_m \frac{h_w^2}{2} \text{ (kN/m)} \quad (1)$$

$$P_2 = K_a y_m h_w (H - h_w) \text{ (kN/m)} \quad (2)$$

$$p_2 = K_a y_b (H - h_w) \text{ (kN/m}^2\text{)}$$

$$P_3 = K_a y_b \left(\frac{H - h_w}{2}\right) \text{ (kN/m)} \quad (3)$$

$$p_3 = y_w (H - h_w) \text{ (kN/m}^2\text{)}$$

$$P_4 = y_w \left(\frac{H - h_w}{2}\right) \text{ (kN/m)} \quad (4)$$

Hence the total active pressure

$$= P_1 + P_2 + P_3 + P_4$$

7.3 Passive pressures on a vertical wall with cohesionless soil backing

Horizontal ground (Figures 7.6 and 7.7)

The intensity of passive resistance at any depth below the horizontal ground surface is given by the equation

$$p_p = K_p y_d \sec \delta \text{ (kN/m}^2\text{)}$$

The total passive resistance

$$P_p - K_p \frac{yd^2}{2} \sec \delta \text{ (kN/m)}$$

As before, Rankine's formula assumes that $\delta = 0$: hence

$$p_p = K_p y d \text{ (kN/m}^2\text{)}$$

$$P_p = \frac{K_p y D^2}{2} \text{ (kN/m)}$$

$$K_p = \frac{(1 + \sin \phi)}{(1 - \sin \phi)} = \tan^2\left(45 + \frac{\phi}{2}\right)$$

Typical values of K_p are given in Table 7.4.

Table 7.4 Values of K_p for cohesionless materials (vertical walls and horizontal ground)

Values of γ	*Values of φ* 25°	30°	35°	40°
	Values of K_p			
0°	2·5	3·0	3·7	4·6
10°	3·1	4·0	4·8	6·5
20°	3·7	4·9	6·0	8·8
30°	–	5·8	7·3	11·4

Saturated soils: horizontal ground (Figure 7.8)

In this case, using Rankine's formula

$$p_1 = K_p y_m d_w \text{ (kN/m}^2\text{)}$$

$$P_1 = K_p y_m \frac{d_w^2}{2} \text{ (kN/m)} \quad (1)$$

$$P_2 = K_p y_m d_w (D - d_w) \text{ (kN/m)} \quad (2)$$

$$p_2 = K_p y_b (D - d_w) \text{ (kN/m}^2\text{)}$$

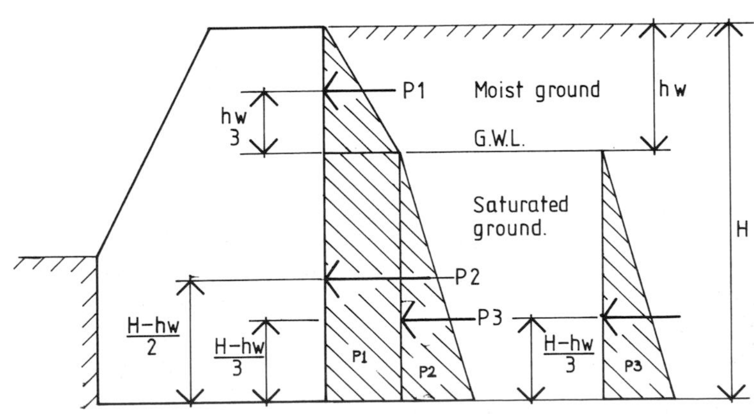

Figure 7.5 Active pressure diagram (Rankine) for saturated ground

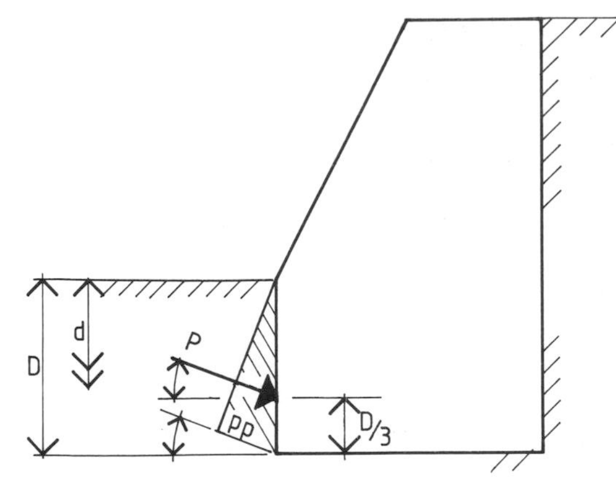

Figure 7.6 Passive pressure diagram (general case)

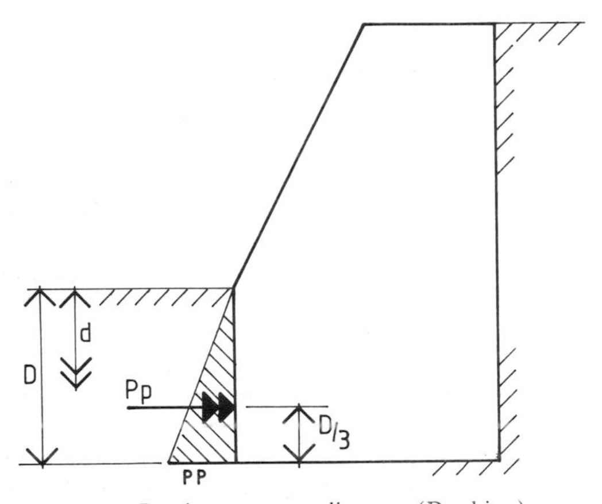

Figure 7.7 Passive pressure diagram (Rankine)

$$P_3 = K_p y_b \left(\frac{D - d_w}{2}\right)^2 \text{(kN/m)} \qquad (3)$$

$$p_3 = y_w (D - d_w) \text{ (kN/m}^2\text{)}$$

$$P_4 = y_w \left(\frac{D - d_w}{2}\right) \text{(kN/m}^2\text{)} \qquad (4)$$

Hence the total passive resistance

$$= P_1 + P_2 + P_3 + P_4$$

7.4 Bearing pressures on foundation soil

The bearing pressures on the foundation soil beneath the base of the wall may be calculated as follows:

1. Where R_v falls inside the middle third of the base:

$$f_1 = \frac{R_v}{B}\left(1 + \frac{6_e}{B}\right)$$

$$f_2 = \frac{R_v}{B}\left(1 - \frac{6_e}{B}\right)$$

2. Where R_v falls at the middle third of the base:

$$f_1 = \frac{2R_v}{B}$$

$$f_2 = 0$$

3. Where R_v falls outside the middle third of the base:

$$f_1 = \frac{2}{3}\frac{R_v}{b}$$

$$f_2 = 0$$

The assumed distribution of bearing pressure for

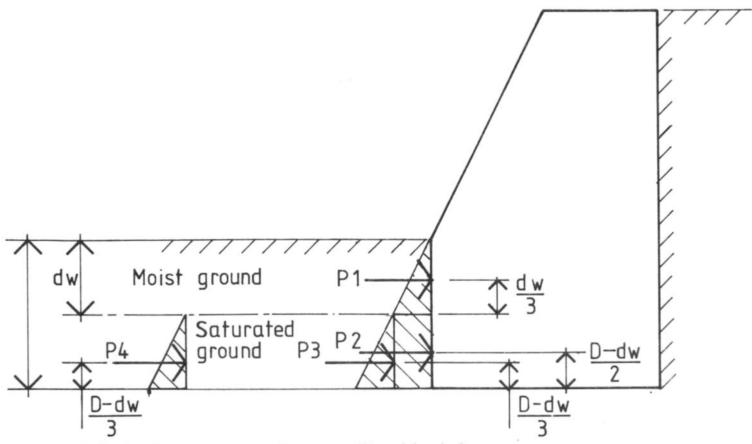

Figure 7.8 Active pressure diagram (Rankine) for saturated ground

Table 7.5 Maximum safe bearing capacities for foundations of width greater than 1 m and depth not less than 600 mm (cohesionless soils)

Type of soil	*Typical values of* ϕ *(degrees)*	*Max. safe bearing capacity (kN/m²)*	
		Dry	*Submerged*
Compact well-graded sands and gravel–sand mixtures	40–45	400–600	200–300
Loose well-graded sands and gravel–sand mixtures	35–40	200–400	100–200
Compact uniform sands	35–40	200–400	100–200
Loose uniform sands	30–35	100–200	50–100

Dry means that the ground water level is at a depth not less than the width B of the foundation below the bottom of the foundation.

the above cases are shown in Figure 7.9. The maximum pressure in all cases should not exceed the maximum safe bearing capacity of the soil. Typical values for cohesionless soils are given in Table 7.5. Note that in gravity walls R_v should not fall outside the middle third of the base.

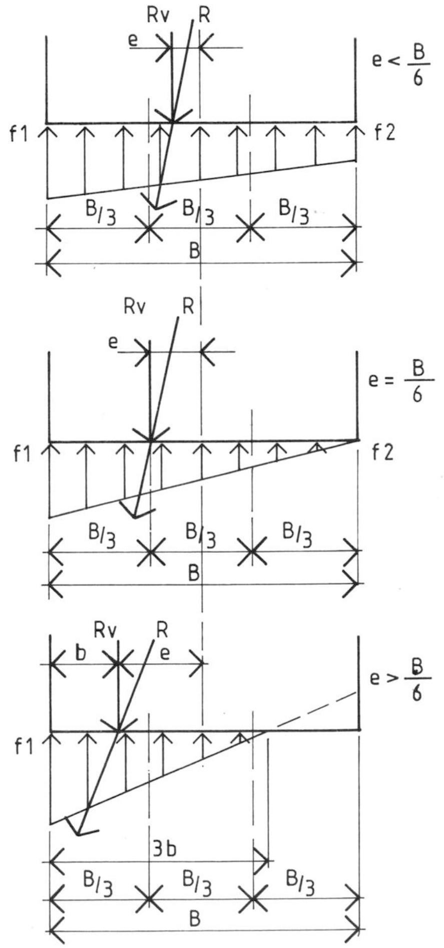

Figure 7.9 Bearing pressure diagrams

7.5 Design of a gravity wall with cohesionless soil backing and horizontal ground, $\delta = 0$

1. Assume a cross section for the wall (the width of the base is approximately half the height of the wall).
2. Calculate the vertical line in which the centroid lies by taking moments of areas about the back of the wall (Figure 7.10).
3. Calculate the weight of one metre run of wall w and the total active pressure P_a due to the earth backing.
4. Find graphically or by calculation where the resultant thrust R of P_a and W cuts the base of the wall and determine the eccentricity e. Graphically, and with reference to Figure 7.11, to scale Oa represents P_a, Ob represents W and Oc is the resultant thrust R. Hence e can

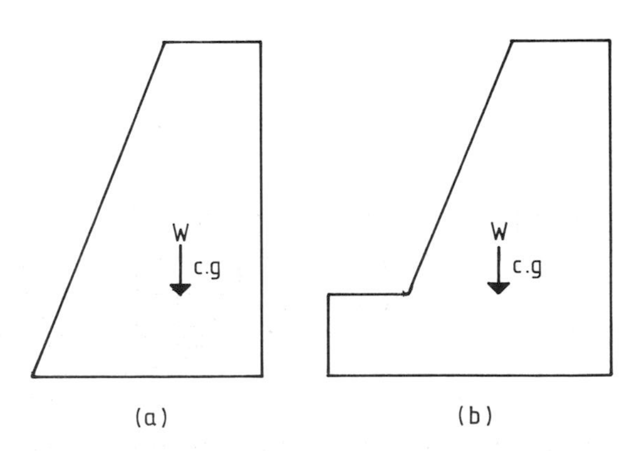

Figure 7.10 The centre of gravity of retaining walls

be obtained by scaling the distance from the midpoint of the base to the point where the resultant cuts the base of the wall. By calculation:

$$x = \frac{P_a H}{3W} \quad e = x - u$$

where u = distance from the centre of gravity of the wall to the midpoint of the base.

5. Calculate the bearing pressure on the foundation soil beneath the base (see Section 7.4).

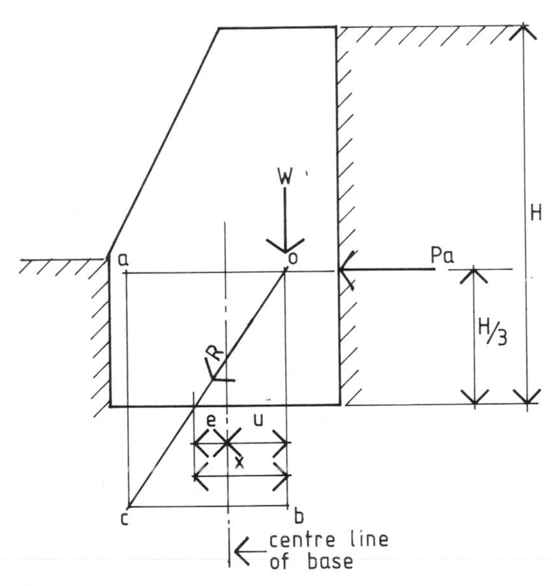

Figure 7.11 Position of resultant thrust in a cavity wall

7.6 Design of a reinforced concrete cantilever wall with cohesionless soil backing and horizontal ground, $\delta = 0$

1. Assume dimensions for the base and stem. The thickness of the base should be at least equal to the thickness of the stem. The width B of the base is about two-thirds the height of the wall for Figure 7.12(a) and about half the height of the wall for Figure 7.12(b) and (c).
2. Calculate the total ultimate active pressure P_a due to earth backing. Note that the pressure acts horizontally ($\delta = 0$) in the case of reinforced concrete retaining walls.
3. Calculate the vertical line in which the centroid lies by taking moments of areas about the heel of the wall (Figure 7.13):

   W_S = weight of stem per metre run (kN),
   W_B = weight of base per metre run (kN),
   W_E = weight of earth per metre run (kN),

   Taking moments about the heel

   $$(W_S + W_B + W_E)\bar{x} = W_S x_S + W_B x_B + W_E x_E$$

4. Calculate the weight of one metre run of wall. Hence

   $$W = W_S + W_B + W_E$$

5. Find where the resultant thrust R of P_a and W cuts the base (as for gravity walls).
6. Calculate the bearing pressures on the foundation soil beneath the base (see Section 7.4).
7. Calculate the ultimate bending moment in the stem (Figure 7.14) and determine the amount of reinforcement required in accordance with BS 8110.
8. Calculate the ultimate bending moments in the

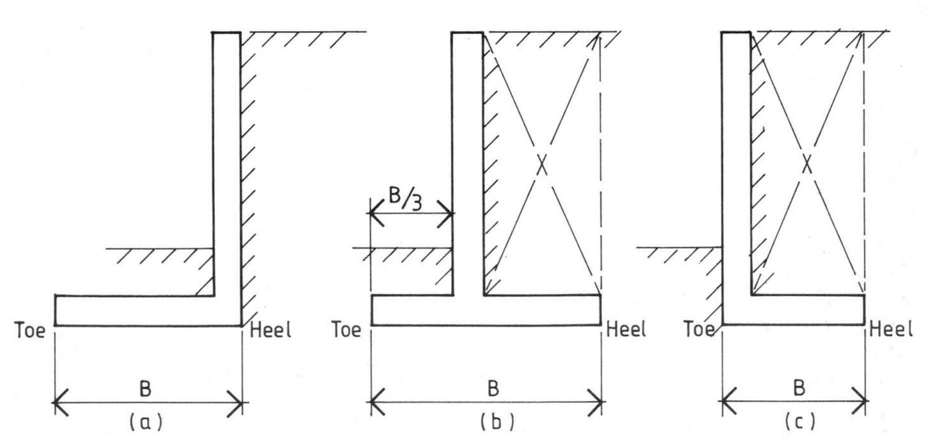

Figure 7.12 RC cantilever wall types

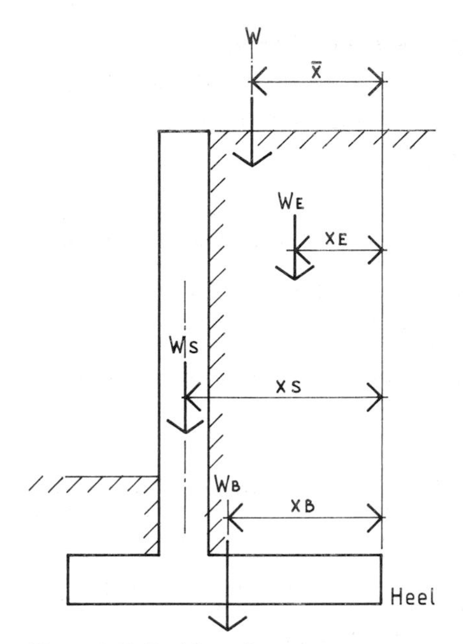

Figure 7.13 Position of resultant vertical load in a reinforced concrete cantilever wall

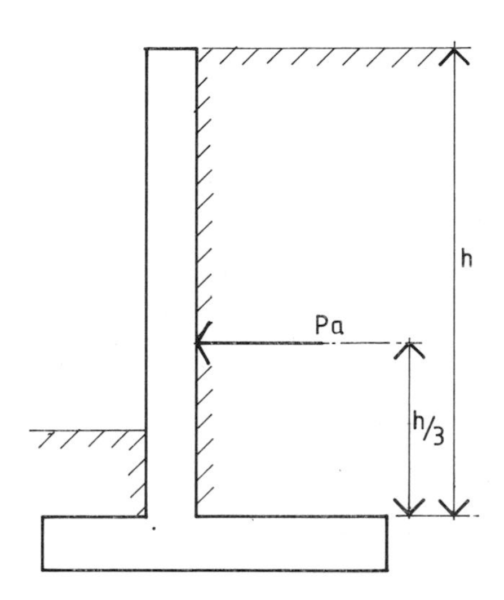

Figure 7.14 Moment at bottom of stem of a RC retaining wall

base. For example, the ultimate bending moment XX in Figure 7.15 is due to downward pressure of earth, downward pressure due to weight of base and upward pressure due to reaction from the soil. Calculate the reinforcement required. Moment and reinforcement calculations should be in accordance with BS 8110.

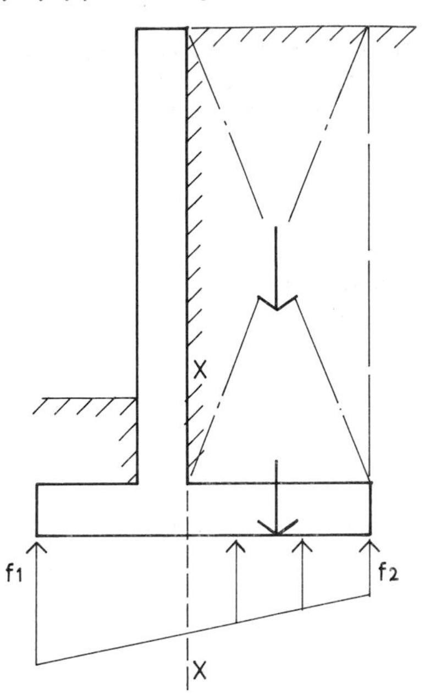

Figure 7.15 Moment in base of a reinforced concrete retaining wall

7.7 Factors of safety for retaining walls in cohesionless materials

Resistance to overturning (Figure 7.16)

In gravity walls the resultant thrust should not fall outside the middle third of the base of the wall. In all other cases a factor of safety of at least 2 is required.

Bearing pressures

The maximum pressure exerted on the foundation soil beneath the wall base, which will usually be at the toe, should not exceed the safe bearing capacity appropriate to the foundation soil. For typical values of safe bearing capacities see Table 7.5. These give a factor of safety of at least 2.

Resistance to forward movement (sliding) (Figure 7.17)

For cohesionless soils, the base friction resistance under a concrete foundation cast *in situ* may be determined by assuming that the angle of friction beneath the base is equal to the tangent of the angle of internal friction of the soil beneath the foundation (see Table 7.1). When the foundation is not cast *in situ* the angle of friction should be taken as equal to δ, the angle of wall friction.

In theory the passive pressure will assist in

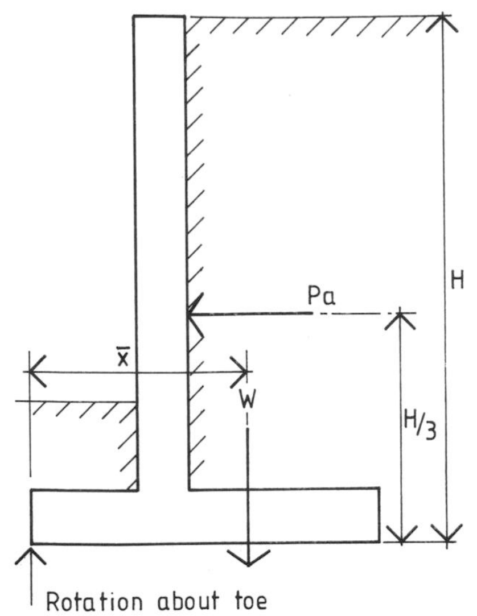

Figure 7.16 Resistance to overturning

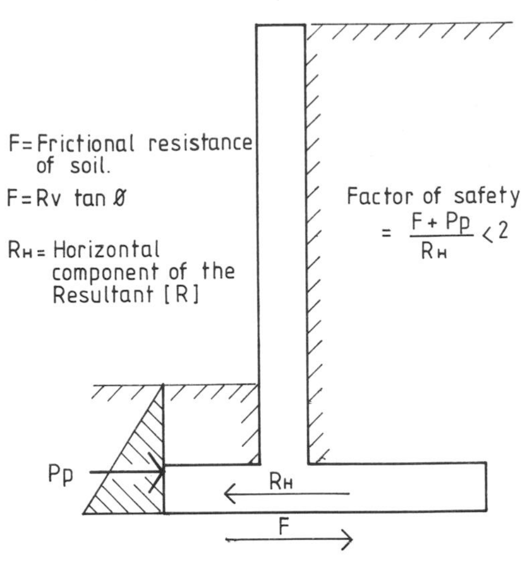

Figure 7.17 Resistance to forward movement

resisting the forward moment of the wall, due to active pressure. This should be treated with extreme caution as the passive resistance of the ground is often sought at relatively shallow depths where the soil is subject to seasonal change.

When computing the total force resisting sliding, base friction or adhesion may be added to the passive resistance of the ground in front of the toe. A factor of safety of approximately 2 should be applied to the total force calculations.

Index